America's Conservation Impulse

Center Books in Natural History
George F. Thompson, series founder and director

America's Conservation Impulse

A Century of Saving Trees in the Old Line State

Geoffrey L. Buckley

The Center for American Places at Columbia College Chicago

The Center for American Places at Columbia College Chicago
600 South Michigan Avenue
Chicago, Illinois 60605-1996, U.S.A.
www.americanplaces.org

18 17 16 15 14 13 12 11 10   1 2 3 4 5

ISBN: 978-1-935195-03-0

Library of Congress Cataloging-in-Publication Data

Buckley, Geoffrey L., 1965–
  America's conservation impulse : a century of saving trees in the Old Line State / by Geoffrey L.
Buckley. — 1st ed.
    p. cm. — (Center books in natural history ; 5)
  Includes bibliographical references and index.
  ISBN 978-1-935195-03-0 (alk. paper)
  1. Forest conservation—Maryland—History. 2. Conservation of natural resources—Mary-
land—History. 3. Nature conservation—Maryland—History. 4. Forests and forestry—Mary-
land—History. 5. Maryland—Environmental conditions. 6. Urban forestry—Maryland—Baltimore—History. 7. Trees in cities—Maryland—Baltimore—History. 8. Baltimore Region
(Md—Environmental conditions. I. Title. II. Series.

SD413.M35B83 2010
333.75'1609752—dc22
                                                                            2010014594

Frontispiece: "US Forest Service, Loblolly Pine, Dr. DeCoursey's place near Carmichael."
Fred Besley's enthusiasm for "champion" trees attracted regional and national attention that
continues to the present day.

To Wells Bradford "Brad" Kormann (1953–2005),
brother-in-law and friend

To see a World in a Grain of Sand
And a Heaven in a Wild Flower,
Hold Infinity in the palm of your hand
And Eternity in an hour.

—William Blake

Contents

Introduction

A Case Study for the Nation

On April 5, 2006, Marylanders observed the 100th anniversary of the Forestry Conservation Act: the law that introduced scientific forest management to the Old Line State. Soon after it was approved in 1906, Maryland became only the third state in the union to hire a trained forester and initiate statewide forest management, placing it at the forefront of the state forestry movement nationwide. Between 1906 and 1942, state forestry in Maryland took root and flourished, thanks in large part to the innovation and leadership of Fred Wilson Besley, Maryland's first—and only—state forester during this period. In an era before ecological forestry was widely practiced or public participation welcomed, Besley, schooled in the precepts of utilitarian conservation at Yale University, formulated a strategy for forest recovery and management that closely mirrored that of his famous mentor Gifford Pinchot, who founded the U.S. Forest Service.[1] Although half a century has passed since Besley's death, he is still remembered for his pioneering work in conservation and sustainable forestry.

Just one week earlier, Baltimore parks and planning officials announced plans to make Baltimore's appearance "softer, greener and more pleasant by doubling the city's tree canopy—the total area covered by leaves—in the next 30 years."[2] Although less well known than Maryland's program, Baltimore's experience with professional forestry dates back almost as far. By some measures it even predates it. In 1912, Baltimore Mayor James H. Preston approved passage of Ordinance No. 154, which created the position of "city forester" and formally introduced professional forest management to the state's largest, and the nation's seventh largest, city.[3] Though not the first municipality in the United States to exert control over street tree planting, Baltimore could now be added to the growing list of major American cities to venture down this road.[4] Despite the best efforts of several professionally trained foresters, Baltimore's aspiration to become known as the "city of a million trees" was never fully realized. Truth be told, the city's "shrinking forest" has been a subject of concern in recent years.[5]

Since the 1970s, a great deal of attention has been devoted to the study of America's public lands. While much of this work has focused specifically on the creation and management of the nation's national parks, forests, monuments, and wildlife refuges, an impressive body of literature has also been assembled on state and city parks. Comparatively little space, however, has been dedicated to exploring the origins of forestry programs at the state and municipal levels.[6] With respect to state forestry, the neglect is surprising given the fertile ground for research that exists, especially in areas situated east of the Mississippi River, where the influence of the U.S. Forest Service is diminished compared to the West and the management role of the states correspondingly amplified.

An examination of state forestry is particularly valuable as it casts light on the tension that developed between promoters of national forests, on the one hand, and advocates of states' rights, on the other. This issue is significant not only because of the lasting effect it has had on the management of Maryland's public lands, but also because Gifford Pinchot and Fred Besley came to stand on opposite sides of the debate. Likewise, the burgeoning interest in urban environments, including the role that trees play in making our cities and towns more livable, suggests that a longitudinal examination of one of America's long-standing urban forestry programs is long overdue. Furthermore, Baltimore's current status as one of the National Science Foundation's two urban long-term ecological research sites (the other is Phoenix, Arizona), its ambitious plan to double its tree canopy in the next thirty years, and its remarkably different experience with professional forestry underscore its value as a study area.

Relying on a rich set of historical sources, including state and municipal government documents, newspaper reports, personal correspondence, and a variety of other archival and visual records, this book offers readers a detailed account of the formative years of professional forestry in Maryland and in Baltimore, from the Progressive Era roots to World War II and into the twenty-first century. One of the first states to hire a state forester and embrace the principles of scientific forest management, Maryland serves as a useful yardstick against which other states' progress may be measured, particularly those in the South, where conservation was delayed. Much the same can be said of Baltimore's forestry program, which is fast approaching the century mark.

About the Book

Some of my fondest childhood memories involve trees. Growing up in Chevy Chase, Maryland, our house was surrounded by big trees: towering white oaks, enormous hickories, oversized hollies, and one very old tulip poplar. One of my favorite pastimes was to sit on the porch with my father when a big thunderstorm blew in. For a young boy, it was exhilarating and terrifying at the same time. The sky would darken, and the wind would blow; lightning would flash, and thunder would crackle. Then the torrential rain would come. The huge trees would bend but not break. When it was over, the sun would reappear, and peace would be restored. If we were lucky, we might also enjoy a break from the humidity.

On one memorable occasion, one of my favorite trees did keel over. It was a massive specimen with four big trunks located in a neighbor's yard across the street. My friends and I loved to climb into the base of the tree and hide from passersby. The tree became, in our minds, a frontier fort, a swift navy frigate, and a World War II fighter plane. We kept enemy British redcoats at bay from behind its secure walls, scanned the horizon for marauding pirates from its bridge, and parachuted from its doomed cockpit. When three of the trunks came crashing down in a storm in the early 1970s, it changed the street forever, at least in the eyes of one disappointed eight-year-old. Deeming the fourth trunk unsafe, the utility company cut it down the next day. When the last remnants of the downed tree were swept away and the tangle of telephone wires and other debris removed from the road, I knew, even then, that the street would never be the same.

Thankfully, there were plenty of other good climbing trees in the neighborhood. The low branches of dogwoods and maples were especially well suited for this purpose, but they were relatively small and presented not much of a challenge. My favorite "climber," on the other hand, had no low branches. The only way to scale it was to hoist myself up the wisteria vine that had long ago attached itself to the trunk. This tree, too, came crashing down during a storm one night—a victim of not only high winds and rain, but also the weight of the heavy vine.

Fortunately, we lived just two blocks from Rock Creek Park in Washington, D.C. Once I was old enough to explore parts of the park on my own, I found a new outlet for my affinity for the woods. When I learned that President Teddy Roosevelt used to venture into the park on his famous point-to-point hikes, it added a new thrill to my periodic visits. Equally exciting to me was the knowledge that American Indians had once roamed these woods, hunting and fishing. That this expansive park was located in the heart of a major metropolitan area did not really register with me at the time. In my mind, it was a wilderness.

Of course, trees are important for other reasons. They offer shade, filter pollutants, and impede soil erosion. A forested watershed plays a critical role in capturing agricultural and storm-water runoff and protecting water quality. Trees also provide food and shelter for wildlife. As far back as I can remember, I have associated trees with wildlife. On warm summer evenings, especially when we had friends over for dinner, my parents would set a tray of chicken bones out on the driveway for the raccoons to eat. Sometimes peanut butter on stale sandwich bread replaced table scraps on the menu. We'd sit on the porch in the dark and wait for our masked guests to arrive. When they had had their fill, we'd watch them scramble back up a nearby tree. In the days before cable television, this was nighttime entertainment at its best.

Then there were the birds. The woodpeckers, from the diminutive downy to the giant pileated, were my favorites. Watching these birds pound away at a snag in search of a meal was truly a sight to behold. "Our" woodpeckers also feasted on slabs of suet my father nailed to the hickory tree at the bottom of the driveway. Occasionally, their quest for food attracted them to the wood siding of our Tudor style house. The early morning hammering worked better than an alarm clock.

Mind you, this was no forest primeval. The whole area was completely denuded at one point. Just four blocks from our home stood a beautiful antebellum farmhouse known locally as "No Gain," a reference, perhaps, to the business fortunes of the property's first owner. As it turns out, the trees I marveled at as a child were relative newcomers on the scene; they were part of a secondary growth forest that gradually reclaimed the old fields of this farm when agriculture was eventually abandoned. It is a sequence of events that was repeated time and again throughout the eastern United States. As fertile land opened up in the Midwest, many farms in the East reverted back to forest. Only later did I come to learn the role that the state played in fostering the recovery of Maryland's forests. And it was later still that I came to appreciate that our cities and suburbs possess forests, too, and that these forests, if they are to thrive, require constant care and maintenance.

My purpose in writing this book is to direct attention to the origins and early history of professional forestry as it evolved at the state and municipal levels in Maryland and Baltimore. In Chapter 1, I review the circumstances that sparked interest in forest conservation at the national level. Happenings on the national stage offer a valuable backdrop against which events taking place at more local scales may be understood more clearly. In Chapter 2, I chronicle the early years of state forestry in Maryland, from its inception in 1906 to agency reorganization in 1923. It was during these years that Fred Besley, a former school teacher from rural northern Virginia and a 1904 graduate of the Yale School of Forestry, devised a strategy for forest recovery in Maryland—one that focused on fighting fires, reforesting "wastelands," establishing a system of public lands, beautifying roadsides, educating the public, and providing technical advice to private landowners. In Chapter 3, I discuss forest management activities as they played out in Baltimore, from early-nineteenth-century efforts to plant trees along the city's streets and public walkways to the first three decades of institutional forest management. I pick up the trail of state forestry again in Chapter 4, following the course of events from 1923 to Besley's retirement nineteen years later. From the perspective of resource management, this was a particularly significant period, for it was during this time that Maryland greatly expanded its state forest and park holdings. In Chapter 5, I bring the stories of Maryland and Baltimore's forests together, placing particular emphasis on the period from World War II to the 1960s. It was during this interval that Maryland—under the watchful eye of Joseph Kaylor and, later, Spencer Ellis—began to invest heavily in the development of the state's parks and recreational facilities, and the city of Baltimore found itself, once again, confronted with a "tree problem." In Chapter 6, I focus on themes of recent trends and enduring legacies. Relying primarily on interviews with resource managers past and present, I draw attention to the challenges—and opportunities—that confront Maryland's—and, by extension, America's—next generation of forestry professionals. I conclude with an epilogue that reflects on the life and work of Maryland's first state forester and Baltimore's first city forester.

In many ways, like the nation, Maryland and Baltimore are at a crossroads when it comes to management of their respective forests. Rampant development, disease, and neglect are among the many problems that threaten the long-term health and stability of these remarkable ecosystems. It is my sincere hope that, by recounting the successes and failures of the past, I will be providing present-day policy makers and resource managers with the critical insights they need to plan wisely for the future. Even more important, I hope this book inspires Marylanders of all walks of life—urban dwellers, suburbanites, and rural denizens alike—to appreciate more fully the beauty and value of the nation's forests. It was not that long ago that Maryland's forests were in a wretched state, ravaged annually by fires, over-exploited by the timber industry, and poorly managed. While much has improved over the years, much remains to be done if this rich natural inheritance is to be enjoyed for generations to come. It is to the story of the forests of the Old Line State that we now turn our attention.

About the Photographs

While many of the unlabeled forestry photographs in the Maryland State Archives collection were likely taken by State Forester F. W. (Fred Wilson) Besley, I have opted not to identify him as the photographer in the *Photography Credits* section unless the image is clearly marked as having been taken by him. And, where appropriate, I include in the captions the original notations made by Besley and other photographers in "quotation marks" without any editorial clean-up or correction.

To drive home the importance of forest conservation, Besley and his staff took thousands of photographs to illustrate the many and varied ways that wood was consumed by Maryland's forest products industries. Reproduced in state forestry publications and converted to glass lantern slides, images such as the ones featured in Figs. 1.3–1.8 were used to educate the general public. Besley also learned and implemented the art and craft of making a professional photograph, even if his intent was documentation not art. The work of noted photographer A. Aubrey Bodine also appears in Figs. 4.28, 4.30, 4.32, 4.33, 4.34, 4.36, and 4.37.

Besley's use of and emphasis on photography for the benefit of public education about forest conservation was pioneering. In later years, organizations such as the Sierra Club relied heavily on the black-and-white photographs of Ansel Adams and the four-color photographs of Eliot Porter to achieve success in saving many national parks and wilderness areas from wanton development and destruction. Unknowingly, Adams and Porter were following in the footsteps of another conservationist, Fred Besley, who became known as "dean of the State Foresters in the United States."[7]

Acknowledgments

Even now, I vividly recall my first encounter with professional forestry in Maryland. It was 1994, and I was a first-year Ph.D. student studying the impacts of coal mining on the forest and water resources of western Maryland. Perusing the open stacks at the University of Maryland's Theodore R. McKeldin Library, I came across the reports of Fred W. Besley, William Bullock Clark, and George B. Sudworth. Their early descriptions of forest destruction left a lasting impression, to say the least. The discovery of this body of work eventually opened up a new avenue of research—one that led to the publication of this book. More important, however, it gave me the chance to interact with remarkable public servants, such as Robert Bailey, Marion Bedingfield, Calvin Buikema, Anne Draddy, Rebecca Feldberg, Offutt Johnson, Ross Kimmel, Jim Mallow, and Francis "Champ" Zumbrun. It is thanks to them—and numerous other forest and park stewards to whom the mantle of conservation has been passed—that Maryland and Baltimore possess the forest resources they do today. I am particularly indebted to Calvin, Offutt, Ross, and Champ, whose goodwill and assistance over many years made this book possible.

Not long after I graduated from the University of Maryland, Morgan Grove of the USDA Forest Service invited me to join the Baltimore Ecosystem Study, a long-term urban ecological research project I have been affiliated with ever since. Were it not for Morgan's strong encouragement and sage advice, this manuscript might never have seen the light of day. Thanks, also, go to BES colleagues Bill Burch, Jackie Carrera, Mike Galvin, Guy Hager, Charlie Lord, Steward Pickett, Mike Ratcliffe, Austin Troy, and Mary Washington. I am especially grateful to Chris Boone, my good friend and research accomplice in all things having to do with Baltimore, and Ron Foresta, whose critical comments to an earlier draft immensely improved the final product.

At Ohio University, I owe a debt of gratitude to Howard Dewald, Leslie Flemming, Ronald Isaac, and Benjamin Ogles, for granting me a faculty fellowship leave during the 2005–2006 school year. I also acknowledge the financial support of the National Science Foundation, the USDA Forest Service, the Maryland Forests Association, and Besley-Rodgers, Inc. Without their backing, I would have been unable to complete this project in a timely fashion.

Numerous librarians and archivists assisted me with the task of gathering data. First and foremost, I thank Carla Heister, Forestry and Environmental Studies Librarian at Yale University, and Francis O'Neill, of the Maryland Historical Society, for their patience and expert advice. I also thank Ed Papenfuse and Rob Shoeberlein at the Maryland State Archives, Chris Becker, Elizabeth Proffen, and Jenny Ferretti with the Maryland Historical Society, and Tom Hollowak at the University of Baltimore's Langsdale Library, for assistance in securing historical photographs. The wonderful staff at several other research facilities deserve recognition as well, including the Baltimore City Archives; the Baltimore Legislative Library; the Enoch Pratt Free Library; Hornbake and McKeldin libraries at the University of Maryland, College Park, especially Elizabeth A. McAllister; the Albin O. Kuhn Library at the University of Maryland Baltimore County; the Vernon R. Alden Library at Ohio University; and the National Archives at College Park, Maryland. Last, but not least, I express my sincerest gratitude to Betty Knupp, Bev Onslow, Helen "Holly" Overington, John Overington, Margaret Rak, Kirk Rodgers, Mary Rotz, Peggy Weller, and Jane Wright, for their willingness to share family memories, letters, and photographs. Kirk's generosity and support, in particular, helped bring this project to a successful conclusion.

Finally, I thank George F. Thompson, founder and director of the Center for American Places, as well as Erin F. Fearing, executive assistant, David Skolkin, book designer and art director, and the other members of the Center's team in Chicago, for making what was once a dream into a reality, and Alexandra, Ingrid, Peter, and Owen, for their inspiration and love.

America's Conservation Impulse

Chapter One

America's Forest Legacy

Get timber by hook or by crook, get it quick and cut it quick—that was the
rule of the citizen. Get rid of it quick—that was the rule of the Government
for the vast timberlands it still controlled. And it has been got rid of both
by the Government and by private owners with amazing efficiency and
startling speed.

—Gifford Pinchot

It would be difficult to find a region in which the useful timber has been
more generally removed than in this county.

—George Sudworth

Conservation history in the United States is generally divided into three over-
lapping phases of federal land policy: acquisition, conveyance, and conservation.
Between 1781 and 1867, the United States acquired approximately two billion
acres of land through treaty, purchase, annexation, and cession (from the original
thirteen states). Intent on settling these newly acquired lands as quickly as possible and keenly interested in economic growth, the federal government began
conveying lands to interested parties almost immediately. Extending from the
late eighteenth century right through to the early twentieth century, the federal
government distributed lands to states, settlers, railroad corporations, and others and offered incentives to promote the development of natural resources. By
the turn of the twentieth century, more than one billion acres of land had been
transferred from public to private hands. By the late nineteenth century, resource
depletion and environmental damage had caused state and federal governments
to begin to replace land-use incentives with land-use restrictions.[1]

The key mechanism driving the transfer of land was the Land Ordinance of 1785, an act which "provided for the orderly conversion of this vast public domain into private property" via a rectangular survey of public lands and periodic auctions. According to Carolyn Merchant, "constant conflict occurred over the minimum amount that had to be bought, and the minimum price that had to be paid." Despite opposition by eastern businessmen "averse to the westward migration of cheap labor," politicians "anxious to raise revenues for paying off Revolutionary War debts," and wealthy land speculators, "the tide turned in favor of small purchasers" with the election of Thomas Jefferson as president in 1800.[2] Successive amendments reduced the minimum acreage that could be acquired from 640 acres in 1785 to 320 acres in 1800 and 160 acres in 1804. In time, the minimum would be reduced to forty acres.

Perhaps the best known of the aforementioned government incentives was the Homestead Act of 1862, which allowed any head of household or anyone over twenty-one to obtain 160 acres of land free of charge if the land was settled and cultivated for five years. Other acts specifically encouraged the mining of minerals and the harvesting of timber. The Free Timber Act of 1878 permitted settlers to cut timber on lands set aside for mineral use. The Timber and Stone Act, also passed in 1878, made 160 acres of land "valuable for minerals and stone as well as timber" available for purchase. These two acts in particular, Merchant informs us, "were bonanzas for timber companies, who hired people to stake claims to forest lands and then turn the lands over to the lumber company."[3] Taken together these acts, along with numerous others passed during the conveyance phase, fueled the rapid deforestation that took place from coast to coast during the nineteenth century.

How extensive was the forest destruction? By 1850, an estimated 113,740,000 acres of land, mostly in the eastern half of the nation, had been cleared for agriculture or otherwise "improved." Between 1850 and 1859, Americans cleared an additional 39,705,000 acres, a remarkable figure when one considers that it was roughly equivalent to one-third of all wood that had been cut during the preceding 200 years. Although the pace of cutting slowed during the Civil War years, it picked up again after the war, with approximately 49,370,000 acres cleared between 1870 and 1879.[4] With respect to the lumber industry, from 1850 to 1910

Fig. 1.1. Deforestation on Meadow Mountain in Maryland's Appalachian region, ca. 1910. Although much of the state was forested at the time of European contact, agricultural clearing and the demands of industry had taken a heavy toll by the end of the nineteenth century.

Fig. 1.2. "Charcoal kiln." In 1850, ninety percent of U.S. energy needs were met by wood. In rural areas, Americans continued to use wood and charcoal well into the twentieth century.

Fig. 1.3. "Tobacco hogsheads Going to market," August 1925.

Fig. 1.4. Approximately forty-nine percent of all wood consumed in Maryland in 1916 was used by the box and crate industry. Most of this wood had to be imported from other states.

Fig. 1.5. "Chestnut poles for transmission lines and method of hauling," July 1925. The American chestnut blight left Maryland with a surplus of chestnut poles.

Fig. 1.6. "Railroad ties at Baltimore," December 1924. Before the widespread use of tar creosote railroad ties had to be replaced every six to seven years.

Fig. 1.7. "Weyerhauser Lbr Co., Curtis Bay. Storage Sheds," August 1926. Curtis Bay (also known as Brooklyn-Curtis Bay) is a neighborhood in a highly industrialized area along Baltimore's waterfront.

Fig. 1.8. "Weyerhauser Lbr Co. at Curtiss Bay. Vessel unloading and storage yds," August 1926.

there was a dramatic increase in production, from 5.4 billion board feet to 44.5 billion board feet per year.[5] Of course, wood was not used merely for construction. Massive amounts of wood were manufactured into railroad crossties, mine props, and fences. Wood was converted to charcoal and burned in iron furnaces. Steam engines, steamboats, and steam locomotives all consumed wood for fuel. Before the widespread adoption of coal, wood was America's principal domestic fuel.[6]

Wood was a versatile resource, but it was not inexhaustible. The increased pace of deforestation—driven by a rapidly industrializing economy, wasteful cutting practices, and improved technology—left large areas practically denuded of trees. Urbanization, especially east of the Mississippi River, also consumed tens of thousands of acres of forested land. Forest fires often followed on the heels of logging operations. Evidence suggests that fires claimed twenty to fifty million acres of forest per year ca. 1900 (an area equal to Delaware, Maryland, Virginia, and West Virginia combined). Deforested and burned-over lands were particularly susceptible to the ravages of floods and droughts, resulting in soil erosion, stream siltation, and wildlife depletion.[7] These changes ultimately led to the "conservation" stage in federal land policy, a time when the federal government, caught up in the current of the Progressive Era, replaced incentives for land-use intensification with a suite of land-use restrictions.[8]

Forests and Forestry

Scholars of American history often refer to the period from the 1890s to the 1920s as the Progressive Era, a time of broad social and economic reform. It was during these years that Jane Addams started the settlement house movement in Chicago, Lewis Hine documented the horrors of child labor, and Jacob Riis chronicled the suffering of America's urban poor. Muckraking reporters, Lincoln Steffens and Ida Tarbell the best known among them, challenged corrupt political bosses and exposed corporate greed. In 1906, Upton Sinclair's novel, *The Jungle*, shocked millions of Americans with its graphic portrayal of life and work in the meatpacking plants of Chicago.

In response to the perceived crisis, the U.S. Congress passed a series of new laws intended to improve conditions for America's working poor and to regulate big business. When Theodore Roosevelt occupied the White House, the Meat Inspection Act, Pure Food and Drug Act, and Hepburn Act, which regulated railroads and pipelines, were signed into law. During Woodrow Wilson's tenure in Washington, the Federal Trade Commission was established to restrain the growth of monopolies, and under William Howard Taft, the telephone and telegraph systems were placed under the control of the Interstate Commerce Commission.[9] While the political motivation behind these various reforms—and their long-term effectiveness—is open to debate, it is clear that all three presidents, including the politically conservative Wilson and Taft, maneuvered to claim the mantle of progressive reformer. In reality, however, Theodore Roosevelt's policies, more than those prescribed by the other two men, laid the groundwork for "the industrial and social service state that came into being under another President Roosevelt later in the century."[10]

The Progressive Era was also characterized by concern for the physical environment. While conservationists such as John Muir subscribed to the idea that nature has an inherent right to exist—a belief that contributed to the creation of a national park system—most progressive reformers viewed nature as "a collection of resources waiting to be used efficiently." Unlike other reform efforts during this period, which were associated with the middle- and upper-classes, the American conservation movement was spearheaded by "a triumvirate of scientific professionals, government bureaucrats, and businessmen involved in resource extraction" that "directed the reform effort from above."[11] According to Samuel P. Hays, "Conservationists were led by people who promoted the 'rational' use of resources with a focus on efficiency, planning for future use, and the application of expertise to broad national problems. But they also promoted a system of decision-making consistent with that spirit, a process by which the expert would decide in terms of the most efficient dovetailing of all competing resource users according to criteria which were considered to be objective, rational, and above the give-and-take of political conflict."[12] Of particular concern to conservationists early on was the condition and availability of the nation's forest resources.

Long before Progressive Era reformers "popularized conservation," fear of an impending "timber famine" prompted many Americans to act.[13] In 1872, J. Sterling Morton of the Nebraska Board of Agriculture launched the nation's first Arbor Day. At the inaugural event, Nebraskans planted more than one million trees across the state, earning Nebraska the nickname, "The Tree Planters

Fig. 1.9. "Tree no. 11–'Gen. Smallwood Chapter'–Arbor Day Planting, 1922." The inscription on the photograph reads: "The Mayor lends a hand, 'for good luck'–Druid Hill Park."

State." [14] Excitement quickly spread to other parts of the country, and, as a result, Arbor Day became an annual celebration in which citizens participated in tree planting activities in cities, suburbs, and rural areas alike. [15]

Fear of a timber famine later convinced the federal government that some sort of corrective action was needed to avert a crisis. After many fits and starts, the U.S. Congress finally resolved that some lands should remain in public ownership. In 1891, President Benjamin Harrison created America's first forest reserve on land adjacent to Yellowstone National Park in Wyoming. [16] Over the next several years, both Harrison and his successor in office, Grover Cleveland, set aside nearly forty million acres of timber reserves, all of it carved from the public domain. In 1897, President William McKinley signed the National Forest Management Act, better known as the Organic Act, which determined the purpose of the new reserves: " to improve and protect the forests" from forest fires and commercial exploitation, to "secure favorable conditions of water flow," and to "furnish a continuous supply of timber for the use and necessities of the citizens of the United States." In 1905, the forest reserves were transferred from the Department of the Interior to the newly created Forest Service under the Department of Agriculture, where they came under the management purview of Chief Forester Gifford Pinchot, the first "scientifically trained forester from the United States." In 1907, the forest reserves were renamed national forests. At the urging of Pinchot, President Theodore Roosevelt, who occupied the White House from 1901 to 1909, created additional national forests from public lands in the West, bringing the total to 150 million acres by 1906.[17] Passage of the Weeks Act in 1911 opened the door for the federal government to begin the process of building national forests in the East, where most land had fallen into private ownership.[18]

Pinchot's approach to forestry was comprehensive. As Hal K. Rothman points out, Pinchot molded his agency "to promote the idea of wise use—the greatest good for the greatest number through scientific management." In addition to managing the new reserves with an eye toward "making them pay," he stressed the importance of research and extension work, fire prevention, reforestation, and public education. Pinchot's "Use Book" was widely distributed among employees, becoming "the Bible of the Forest Service" and "serving as the text from which foresters made local decisions."[19]

If the Progressive Era is remembered as the springboard for forest conservation in the United States, perhaps it should also be remembered for the brand of conservation it ushered in. To a remarkable degree, Americans at this time placed a tremendous amount of faith and responsibility in the hands of a few select experts whose job it was to identify and solve resource problems. To be sure, the influence of engineers, foresters, landscape architects, and soil scientists, was ascendant in these years. Mark Baker writes that the Progressive Era was a time of "unbridled enthusiasm for science and the ability of technically trained experts working from within scientifically organized and politically neutral bureaucracies to articulate and achieve the public good."[20] The environmental knowledge produced and the conservation strategies employed by these "technically trained experts" went largely unchallenged. Only recently have we come to question the "authoritative role of scientists in producing environmental knowledge" and to probe more thoroughly the social and political context within which neutral and "objective" observations were made.[21]

The Urban Forest

In their book, *City and Environment*, Christopher G. Boone and Ali Modarres call our attention to the fact that trees may be found in abundance in urban as well as rural areas. "Only from the window of an airplane," however, is it made plainly obvious to us that American cities are often "heavily forested," covering, on average, more than a quarter of urban land.[22] Indeed, it is often difficult to determine where the "urban" forest ends and the "non-urban" forest begins. If the political boundaries that separate our urban forests from their more rural counterparts are sometimes difficult to discern, their purpose and the manner in which they are managed often sets urban forests apart. Although urban trees have long been valued for the ecological services they perform, in the past they were planted mainly to fulfill certain social and cultural needs.

In 1985, Jon Peterson called urban open space a "publicly created artifact, deliberately held open to satisfy certain functional requirements of urban life."[23] In some cases, these functional requirements are met by streets and parking lots. In other cases, more "natural" or "green" open spaces are needed to meet the demands of urban residents. These may include town squares, parks, schoolyards, golf courses, community gardens, vacant lots, tracts of privately owned land, and even cemeteries.[24] Many of these spaces support, or are capable of supporting, trees. Collectively, this patchwork quilt of forested spaces in the midst of the built environment, stretching from the urban core to the suburban periphery, comprises what we commonly refer to as the urban forest.

Of course, trees have always been present in the urban landscape. Sometimes they constitute nothing more than residual vegetation. Occasionally, they crop up in unexpected places as opportunistic volunteers. More often than not, however, their presence is more deliberate. "Like buildings," writes Henry W. Lawrence, "urban vegetation has always been part of a set of landscape forms, developed in the context of particular cultural and ecological conditions." According to Lawrence, tree planting in European and North American cities has been carried out largely within the context of four landscape settings: linear promenades, small squares, large parks, and private gardens.[25]

In Europe, the practice of planting trees along streets and waterways and beside or on top of walls dates to the Renaissance. France was particularly

Fig. 1.10. "No. 7 Park Avenue." Baltimore took pride in planting trees along its major thoroughfares and, selectively, along other street trees.

renowned for its tree-lined promenades. After Paris began lining its boulevards with trees, largely for aesthetic reasons, the practice spread to other European cities. Indeed, as Lawrence reveals, by the early years of the nineteenth century most major western European cities possessed "at least one such promenade," while most towns in France "had several of them." Across the Atlantic, the tradition of placing trees along streets "in front of homes and businesses for shade and ornament" dates to the colonial period and represents something uniquely American: "Less formal and less regular than European urban allées, colonial American street trees reflected in their irregular spacing and use of diverse species the individualism that characterized American society in comparison to Europe." The custom of planting trees along streets in urban areas gradually gained momentum in North America. In 1682, William Penn included several tree-filled open spaces in his design for Philadelphia.

In colonial Massachusetts, not only were roadside trees plainly in evidence, but "early laws imposed fines for their mutilation." [26] In Michigan Territory in 1807, legislators endorsed the idea of planting trees along boulevards and in squares in Detroit. In 1821, officials in Mississippi promoted tree planting throughout the state's new capital. [27]

Although public squares and thoroughfares are notable as some of the first places where trees were established as part of the built-up area of cities and towns, by far the largest swaths of the urban forest today are found in America's great urban parks. Starting in the early nineteenth century, trees and other "natural" features were introduced to cities such as London as part of a romantic landscape movement effort to counter the negative effects from the rapid expansion of industrialism. [28] The spread of industrialism in America elicited a similar reaction among conservationists and social reformers, leading to city beautification, the planting of street trees, and park construction. The development of New York City's Central Park, in particular, sparked a movement in the United States to create large new parks in every major American city. [29] As Terence Young points out, many of the urban problems plaguing our cities during the second half of the nineteenth century were thought to be linked to "the alienation of urban residents from nature." Park advocates who subscribed to this rather deterministic view of the physical environment argued that social problems "did not arise because people were evil but because they were out of touch" with nature. "When urban society was once again brought into contact with nature," notes Young, "many of the problems would be alleviated," or so park advocates thought. [30]

Trees have long been valued for the many and varied services they provide to urban dwellers. For centuries, urbanites have considered shade to be a public health asset, especially if it is supplied by a mature forest canopy. By the nineteenth century, trees were also being praised for their ability to filter pollutants. More recently, social scientists have suggested that people suffering from "mental fatigue" may experience some measure of healing when exposed to a "restorative environment" such as an urban forest or park. Not only does the placement of trees along roadways and paths create more pleasant environments by framing attractive landscapes and screening out undesirable ones, but roadside planting has also been shown to relieve driver stress. [31]

The value of the urban forest extends well beyond the realm of public health. Among other things, trees store and sequester carbon dioxide, reduce summertime air temperatures, and mitigate the urban heat island effect. They reduce runoff rates and flooding, impede erosion, improve storm sewer capacity, and provide habitat for plant and animal life.[32] In addition, trees have been shown to have positive social and economic impacts on urban communities. They have been linked to lower crime rates and higher market values for homes and rentals. They also contribute significantly to energy savings, particularly in the summertime months.[33]

Unfortunately, numerous obstacles prevent city residents from taking full advantage of the benefits offered by the urban forest. In her classic volume, *The Granite Garden: Urban Nature and Human Design*, Anne Whiston Spirn alerted us to the challenges resource managers face when charged with care and maintenance of urban trees, especially those that line busy thoroughfares:

> Street trees . . . eke out a marginal existence, their roots cramped between building and street foundations, threaded among water, gas, electric, and telephone lines, and encased in soil as dense and infertile as concrete. Their trunks are gouged by car fenders, bicycle chains, and even the stakes installed to protect them. Their branches are pruned by passing buses. Leaves and bark are baked in the reflected heat from pavement and walls or condemned to perpetual shade cast by adjacent buildings. Roots are parched or drowned by a lack of water or an overabundance; in either case, their ability to deliver essential nutrients to the tree is drastically reduced.[34]

Such conditions reduce the life expectancy of urban trees to as low as eight to ten years.[35] In addition to the problems mentioned above, disease, construction projects, pollution from motor vehicles, inconsistent and insufficient funding, and apathy on the part of property owners threaten the viability and long-term sustainability of the urban forest.[36]

Though long viewed by environmentalists and others as " the ultimate expression of human domination over nature," a growing number of scholars, planners, and decision-makers have come to acknowledge that cities are "fundamentally embedded in a natural environment." [37] Today, we view the city not as the antithesis of nature but as an ecosystem with, among other things, a unique hydrology and distinctive vegetation. Increasingly, we acknowledge the critical functions that the urban forest—or, as Gary Moll states, the "area in and around the places we live that has or can have trees"—carries out in this inimitable environment. [38]

A "Thickly Wooded" Land

Travelers' accounts, settlers' diaries, and surveyors' notes provide valuable information concerning the extent, condition, and diversity of colonial Maryland's forest prior to major agricultural and industrial clearing. Writing in 1633, Father Andrew White described the country he saw as, "for the most part, thickly wooded" with "a great many hickory trees, and oaks so straight and tall that beams, sixty feet long and two and a half feet wide, can be made of them." He also took note of the cypress trees that "grow to a height of 80 feet before they have any branches"; trees so large that "three men with arms extended can barely reach round their trunks." [39] In 1838, Benjamin Silliman, Professor of Chemistry, Metallurgy, and Geology at Yale University, conducted a survey of mineral lands in the mountainous western portion of the state for the presidents of the Maryland Mining Company and the Maryland and New York Iron and Coal Company. Along the way, he recorded his impressions of the region's timber resources:

> The extensive and numerous forests are filled with excellent timber. White and Spanish oak, and other principal varieties of this family—hickory in its leading kinds—the butternut—the beach—the sugar maple—the chestnut—the cucumber tree—white pine, locust, &c. The trees are of full size and in thrifty condition; so that every estate in the territory may be fully supplied with the best timber for buildings, for machinery, and construction of every kind; and should charcoal be needed extensively in the arts and manufactures, that will certainly spring up in a region abounding with the best mineral coal, the mountain forests will, for a long time, continue to afford full supply. [40]

Others documented the existence of tremendous tulip poplars on Maryland's Eastern Shore and towering white pines in its western mountains.

Throughout the eighteenth and well into the nineteenth centuries, visitors to Baltimore commented on the city's trees. One tourist in the 1780s, Luigi Castiglioni, marveled at the trees in the vicinity of the city: "He was much impressed with the forests of black pines, black oaks and some superb tulip trees along the Patapsco River." Frances D'Arusmont, who spent two years in the United States from 1818 to 1820, was very taken with Baltimore. She found the city "spread over three gentle hills; the streets, without sharing the fatiguing regularity of and unvarying similarity of those of Philadelphia, are equally clean, cheerful, and pleasingly ornamented with trees. . . ." Henry Tudor, an English gentleman touring Baltimore in 1831, considered the city to be unusually "rich in natural scenery, and varied by all the requisites of hill and dale, wood and water, verdant meadows and well cultivated lands, with heights embellished by country-houses. . . ." Simon O'Ferrall, another visitor to the city, also brought up the subject of the city's trees in his journal. According to Raphael Semmes, "O'Ferrall . . . commented on the fine trees which lined many Baltimore streets—'Lombardy poplar, locust, and pride-of-china trees—the last mentioned especially afford a fine shade."[41]

Starting in the second half of the nineteenth century, eyewitnesses began to portray a very different picture of Maryland's forests. By this time, agricultural clearing, charcoal production, commercial logging, and the demands of coal mining and railroad companies had stripped away much of the state's forest cover. When the federal Census Bureau undertook a national tree count in 1879, the Tidewater counties on either side of Chesapeake Bay were identified as having particularly low densities of timber.[42] In 1900 and again in 1906, the Maryland Geological Survey, under the direction of State Geologist William Bullock Clark, took the opportunity to comment on forest conditions. With regard to the state's Appalachian region, Clark observed:

> What little virgin forest there is in Maryland is located in inaccessible parts of this region. . . . Nearly all the merchantable coniferous trees have already been culled from the forests . . . and the hardwoods are now rapidly being cleaned out under the highly intensive system of lumbering which has lately been inaugurated in the region. Trees of nearly all species down to very small sizes

Fig. 1.11. "East view of Baltimore, Maryland." Visitors to Baltimore often commented on the city's trees.

Fig. 1.12. "View looking east from Shot Tower, ca. 1850." The extent of Baltimore's urban forest becomes apparent when viewed from above, as in this bird's eye view.

are used for mine props and lagging. The prevailing forest condition is that of cut-over virgin forest, covered with a scattering growth of large, defective trees not suitable for lumber, interspersed with reproduction of hardwood sprouts and seedlings, and occasional patches of coniferous reproduction.[43]

George B. Sudworth, a dendrologist in the Forestry Division of the U.S. Department of Agriculture, rendered another opinion. In reference to Allegany County in western Maryland, Sudworth confessed: "It would be difficult to find a region in which the useful timber has been more generally removed than in this county. . . ."[44]

Baltimore's trees, especially its street trees, also appeared to be in decline at this time. Although tree-planting activities were carried out periodically, the city's trees suffered from poor maintenance and neglect. In its annual report to the mayor and the city council in 1904, the Board of Park Commissioners commented on the "shameful condition" of the city's trees.[45] Whether it was the streets of Baltimore, the rolling hills of the Piedmont, or the mountains of western Maryland, one thing was certain as the nineteenth century drew to a close: forest conservation was an idea whose time had come.

Chapter Two

Professional Forestry Comes to Maryland, 1906–1923

...I believe that there is no body of men who have it in their power to-day to do a greater service to the country than those engaged in the scientific study of, and practical application of, approved methods of forestry for the preservation of the woods of the United States....

—Theodore Roosevelt

Politics do not fight fires. I never ask a man his politics or his religion. Fighting forest fires is a hard job and takes the best men in both parties.

—Abraham Lincoln "Link" Sines

Concern over the loss of forest resources was not limited to the federal government. Indeed, in the opinion of Chief Forester William B. Greeley, the states —not the federal government—were at the forefront of the forest conservation movement. Writing in 1927, Greeley identified several states which had "inaugurated inquiries into their forest conditions," issued "bounty or tax exemption laws to encourage timber planting," assembled committees to investigate forest policy, or launched "forestry bureaus or commissions," all in advance of federal government action. He noted that New York had set up a commission in 1872 to consider state ownership of forestland and that, by 1885, was acquiring land that would form the core of Adirondack State Park. He also singled out Pennsylvania as a state "in the lead of the Federal Government" when it came to "inaugurating a real program of forest ownership."[1] Little wonder that Ralph Widner concluded more than forty years ago that American forest policies were born "not in the capitol in Washington, but out in the states."[2]

Fig. 2.1. "Col. Wm. B. Greeley, Chief of the Forest Service, receiving the spade from Erwin Williams of Michigan preparatory to the planting of the 4-H hemlock tree on the grounds of the Dept. of Agri. at Washington, D.C. Other 4-H Club boys on the planting crew are Laban Ladd of N.H., Alfred J. Naquin of La., & Dan Reaugh of Wash. June 18, 1927." Like Besley, Greeley was an early advocate of leaving public forest management to the states wherever possible.

Although not among the very first states to embrace professional forestry, it was not long before Maryland joined the vanguard. As the twentieth century unfolded, vast stretches of once-forested land in the state lay in ruin. Settlement and agricultural expansion over the preceding 250 years had combined to reduce forest cover across the state dramatically. The demands of a rapidly industrializing economy had likewise taken a heavy toll, altering forest composition and greatly diminishing the availability of valuable commercial species. By 1890, the state no longer produced enough timber to meet its needs.[3] The fate of Maryland's remaining forests hung in the balance.

In 1906, conservationists, outdoor enthusiasts, and state government officials—aroused perhaps by events taking place on the national stage (such as the establishment of national parks and monuments) and encouraged by a generous donation of land—took the first small steps toward addressing the problem. In short order, the Maryland Forestry Conservation Act was drafted and passed, a State Board of Forestry was established, and Fred Wilson Besley was installed as Maryland's first—and only the country's third—state forester.

The challenge he faced was a formidable one. Few people had heard of professional forestry, let alone practiced it. Timber production was down, and fires were rampant. There were no state forests or parks. Given that there were few models upon which to base a forestry program and hampered by an anemic budget, Besley was forced to improvise and innovate. A recent graduate of the Yale School of Forestry and a former student assistant of Gifford Pinchot's, Besley was well qualified to do so. Over the course of a career that spanned thirty-six years, he and his staff worked diligently to introduce professional forestry principles to the Old Line State. Among Besley's many accomplishments, he conducted a painstaking statewide survey of forest resources; implemented an aggressive fire management policy; instituted a program of reforestation; promoted the planting of roadside trees; and, in later years, added significant acreage to the state's system of forest reserves. Through it all, he proved adept at negotiating the shifting and sometimes treacherous economic and political terrain.[4]

From Savage River State Forest in the rugged western part of the state to the expansive Patapsco Valley State Park at Baltimore's doorstep, to the Pocomoke State Forest on the Eastern Shore, the Maryland Forest Service and the Maryland Park Service, together, manage more than 330,000 acres of public lands.[5] The foundation for these public holdings was laid down by Besley during his career. Less noticeable but just as significant, today's forests—both public and private—still bear the mark of Maryland's first state forester. To say that Besley loomed large when it came to the practice of professional forestry in these early years would be an understatement. It is to Besley's pioneering efforts in state forestry that we now turn our attention.

The Garrett Bequest

In 1906, Robert and John Garrett gave three tracts of mountain forest land to the state of Maryland. But there was a string attached. On the occasion of the "Golden Anniversary of Maryland Forestry" in 1956, eighty-year-old Robert

Garrett explained: "I became familiar with forestry in the Adirondack Mountains in New York in the 90's, and realized we should do something about it in Maryland. I got in touch with McCulloch [sic] Brown (State Senator from Garrett County) and told him my brother and I would donate 2,000 acres of land for state forests, if the Legislature would put through a bill to take care of it."[6] According to the deed that was drawn up, should the state of Maryland fail to uphold its end of the bargain, then "title to the said several tracts and parcels of land shall revert to the said donors . . . [who] shall have the right to take over the possession of said tracts of land, and hold them the same as if said gift had not been made."[7]

Eager to take advantage of the Garretts' offer, Brown set about drafting the Maryland Forestry Conservation Act of 1906. While he is often credited with having written the forestry bill, it is almost certain that he had help. In 1913, Besley himself affirmed that "the Maryland forest law . . . contained many desirable features copied from the laws of other states where they were in successful operation." A more intriguing possibility is that Brown received a helping hand from America's leading authority on forestry. Edna Warren suggests as much when she states: "Whether Brown availed himself of assistance from his fellow alumnus, Gifford Pinchot, then head of the Division of Forestry in the U.S. Department of Agriculture, or of the Yale Forestry School, is not known. But he wrote a forestry bill when state forestry was an uncharted wilderness. . . ." With the help of General Joseph B. Seth, a "leading citizen" of Talbot County and President of the Maryland State Senate, he then worked to push the act through the legislature. Passage of the act paved the way for the creation of the Office of the State Forester and the appointment of a State Board of Forestry. And thus the cornerstone was laid for Maryland's current system of public lands.[8]

In retrospect, it is somewhat ironic that the roots of professional forestry in Maryland should be traced to the Garrett brothers and to McCulloh Brown. The land Robert and John Garrett gave to Maryland came from the estate of their grandfather, Baltimore and Ohio (B&O) Railroad magnate John Work Garrett. Perhaps no other industrialist played such a central role in the deforestation of western Maryland as the elder Garrett, for the introduction of the B&O opened the region to large-scale coal mining and logging. Similarly, Brown's grandfather,

James W. McCulloh—lawyer, political operative, and financial deal-maker—was a major speculator and developer during the first half of the nineteenth century. Two generations later, the heirs of these two foremost capitalists sought to replenish the resource upon which their family fortunes were founded. An appropriate closing of a circle, if you will.

"The Young Man You Want is Besley"

In the years leading up to the Civil War, Americans living in the Northeast and Mid-Atlantic pulled up stakes and headed west and south in large numbers in pursuit of new economic opportunities. Bartholomew Besley and Sarah Elizabeth Wilson, both of Ulster County, New York, were swept up in this wave of migration. Along with their parents, Isaac and Ann Besley and William and Deborah Wilson, they migrated in their youth from the Hudson Valley to northern Virginia. For the Besleys, of Huguenot and Quaker descent, and the Wilsons, whose forebears came from England, the transition would not be an easy one.[9]

With little or no opportunity for formal education, Bartholomew and Sarah helped their parents "lumber off the virgin hardwood forest and bring land under cultivation, or nurse back into productivity the old fields long exhausted under cash croppings." When the Civil War broke out, the young couple postponed their wedding, and Bartholomew enlisted with the First Volunteer Cavalry Regiment from New York. Sarah and her mother, along with her sister and young brother, remained behind to look after the farm. Two of her brothers went to war. Four years later, Bartholomew returned home, withered and gaunt, from Andersonville, Georgia, where he had spent the last year of the conflict in the notorious Confederate prison. Sarah's two brothers were killed.[10]

Soon after the war, Bartholomew and Sarah were married. They took up farming and started a family. First came two daughters: Grace Adell in 1868 and Elsie May one year later. On February 16, 1872, Sarah gave birth to twins: Florence Eugenia and Fred Wilson. Two more children, Naomi Inez and George Lamoree, followed in 1876 and 1884, respectively. Sarah and Bartholomew, "robbed of their educational opportunities by their transplanting to the woodlands of Virginia," and, perhaps, motivated by the hardships they endured during the Civil War, sought to ensure that their children carved out for themselves a better life: "The boy should not settle down to work with his father's tools. The girls should push into the new fields opening to women. Greater comfort and relief from the grim struggle at home could wait. The parents would continue as they had begun. The children must go on."[11]

From these humble beginnings, Fred Wilson Besley would one day emerge to break new ground in the field of professional forestry. But this career path was not yet open to him. For all practical purposes, professional forestry did not exist in the America of his youth. Having "been through" all the textbooks available to him at Freedom Hill—the local one-room public school—Besley enrolled at the Maryland State Agricultural College (now the University of Maryland at College Park), "where he studied military tactics and served in the college battalion."[12] Upon graduating in 1892, it was his intention to enter the field of engineering "but it was a time of depression" and "jobs were scarce."[13] Instead, Besley returned to Fairfax County, where he accepted a teaching position in the public school system. In time, he rose to the rank of principal of a village school, the same one he had attended and at which he later taught, serving in that capacity between 1897 and 1900. During the summers, he augmented his income as Deputy Treasurer of Fairfax County.[14]

It was during this period in his life that Besley met Bertha Adeline Simonds. Born in Oneida Castle, New York, on April 3, 1876, Bertha's family moved to Washington, DC, where her father, Elmer Simonds, ran a boarding house. He also owned a dairy farm near Vienna, Virginia. Although she suffered from poor health—she contracted scarlet and typhoid fever as a child—Bertha nevertheless helped raise her five siblings. After completing high school, she moved to Kansas City for two years to live with her father's ailing stepsister, Addie. Vivacious, beautiful, and musically inclined, Bertha had always been socially active. Her extended stay with Aunt "Addie," who had married into the wealthy Armour family, left a lasting impression, exposing her to an entirely different circle of friends and acquaintances. Bertha made her debut first in Kansas City, and, later, before President Grover Cleveland at the White House. When an accident left her father paralyzed from the waist down, Bertha dutifully returned to Virginia to care for him. Unable to manage the farm, Elmer sold it and relocated the family to another property in Vienna. Bertha gradually became accustomed to the slower pace of life in Vienna. It was here, at an adult Sunday school class, that she first met Fred Besley.[15]

According to family sources, Bertha's outgoing and friendly nature stood in marked contrast to Besley's quiet confidence and serious demeanor. She was ebullient and ambitious. He was humble, hardworking, and very religious. She was raised in the city. Her family was socially connected. He was from the country. His father was a farmer. Rather than repel, these differences brought the two young people together: "Each brought to the other glimpses of a slightly different world. Each saw in the other something of which each felt the lack. Each felt for the other admiration tempered with awe. Slowly but surely they were irresistibly drawn together."[16] While friends and family members may have harbored doubts about their compatibility, Bertha Simonds and Fred Besley, secure in the religious faith they shared, were more self-assured. On September 19, 1900, they were married at the Vienna Presbyterian Church. With help from his father, Besley built a home on Leesburg Pike near Tyson's Corners and the newlyweds settled in.

Fred Besley's life then took an unexpected turn. In 1900, he resigned his teaching post and accepted a position working as one of Gifford Pinchot's "student assistants" at the Forestry Division. Hired "for little more than expenses and honoraria" the chief forester utilized this cheap source of labor as a means of stretching his meager budget. He also used this as an opportunity to groom future employees.[17]

The position of student assistant was a highly coveted job. Of the 232 individuals who submitted applications, Pinchot selected just sixty-one, roughly one out of four. While the competition was stiff, the pay was barely enough to live on. Students working in the field were paid $25.00 per month. Statistical work at the headquarters in Washington garnered $40.00 per month. Given Besley's educational training and professional experience to this point and considering the meager pay, one wonders what compelled Besley to shift his career plans so drastically. The answer, it turns out, can be traced to a chance encounter with Pinchot in the halls of the Department of Agriculture two years earlier. According to one account: "The two men . . . met in 1898 when Besley went to the Department of Agriculture to talk about a more lucrative career with Henry Albord, director of the dairy division, who was a friend of the family.

Fig. 2.2. "State foresters at their first meeting in Harrisburg, Pennsylvania, in 1920." This is the only known photograph that includes Fred Besley (back row and third from right) and his mentor, Gifford Pinchot (first row and fifth from left).

Major Albord introduced the dissatisfied young school teacher to Pinchot." Although Besley had never even heard of professional forestry, he later recalled that "Pinchot was so boiling over with enthusiasm about forestry, that then and there I adopted forestry as my career."[18]

Besley spent the next two field seasons traveling to different parts of the country, conducting surveys and acquiring practical field experience. His personal recollections, put to paper many years later, reveal a youthful exuberance and penchant for hard work: "My first field trip was in August, 1900, when I was sent with a crew of five to cruise a tract of private timberland in the Adirondack Mountains, making strip surveys of timber. From there we were sent to make a survey of jack pine woodlands in Michigan to determine their suitability for a state forest. On the return trip to Washington, my ego got the biggest boost I ever experienced. When we were about to miss connections in Detroit, the conductor wired ahead to hold the Washington express for 'a group of federal men.' And the train was held 30 minutes for us—young fellows earning $25 a month."

During the winter months, field crews returned to Washington, where the forestry assistants compiled statistics using the data they had collected. In addition, they were "advised to do some forestry study" by "reading such books on forestry as were available." The highlight of this time spent in the nation's capital was the weekly meeting of the "Baked Apple and Gingerbread Club" at Pinchot's home on Rhode Island Avenue. In addition to the guest lectures on forestry, student assistants were served a hearty meal. "Believe me," recalled Besley decades later, "to young men living in Washington on $40 a month, some of us with families, the highlight of the evening was often that free meal of baked apples and gingerbread with plenty of milk to drink." At one memorable gathering, a special guest encouraged the group to dedicate their lives to conservation: "Although it is an unwritten law that the President of the United States does not address groups in private homes, so greatly did Teddy Roosevelt share Pinchot's enthusiasm for the conservation of natural resources that he broke the rule to urge us to make conservation our life's work."[19]

The return of warm weather saw the student assistants assigned to the field again. For Besley, this meant traveling to eastern Kentucky in the summer of 1901, and it was here that he first encountered suspicion among locals: "We were working in an area where illicit whiskey making [had] been very active and was still practiced. Consequently when we arrived and set up camp as a federal outfit we were suspected as revenue agents using our tree calipers and forest activities as a blind. We were carefully watched from behind trees. It took 2 or 3 weeks to convince them that we were harmless, after which we were treated royally with characteristic mountain hospitality."[20] This was a fortunate turn of events, as Besley reminisced in another account of the same trip: "In the meantime, as it was before the days of canned goods, we nearly starved. When their suspicions were finally allayed, they kept us bountifully supplied with provisions, including squirrels at a nickel apiece for stews."[21] While posted in Kentucky, Besley was sent a camera and instructed to take photographs of trees and forest products. Although he had never used such a device before, he quickly mastered the techniques involved. It was a skill that would serve him well in the years to come.[22]

After a short stint in Washington, Besley was sent into the field again in November, this time to Texas, where his duties included "counting rings on stumps of long leaf pines" on the property of the Kirby Lumber Company.[23] On this trip, Besley appears to have had an epiphany. "Forestry was becoming an established profession, and after 2 years as a sort of apprentice I had come to realize that if I was to make it my profession a technical education was necessary for any real advancement. I had a momentous decision to make. I was married and with a second child just added to the family and with very limited financial resources.... I had made inquiry about entering the Yale Forest School and found that I could be accepted for the 2 year course by entering January 1st and making up for the time lost since the beginning of the school year in September."[24]

At the time, there were few options for Besley. A forestry program had been founded at Cornell University in 1898 with the former Chief of the U.S. Division of Forestry, Bernhard E. Fernow, serving as its dean, and a one-year program had officially existed at the Biltmore Forest School in North Carolina since 1897 under the direction of Alvin Schenck. By 1910, several other states had established forestry schools, including California, Michigan, Pennsylvania, and Washington. But it was Yale University that "led and set the standard" for the other schools. Its influence was such that, when University of Michigan alumnus Lyle F. Watts was appointed Chief of the Forest Service in 1943, he was the first appointee not to have earned a degree from the Yale Forest School. Yale admitted its first students to the Forest School in the fall of 1900 primarily in response to an "endowment gift" from the Pinchot family of $150,000, "an amount that would shortly be doubled."[25] Given Besley's admiration for Pinchot and the fact that the chief forester's family had recently endowed the school, Besley's choice of Yale was an obvious one.

After receiving a release from the field party in Texas, Besley sold the house in Virginia and moved his family to New Haven. Once again, money was in short supply, but they "managed somehow to subsist for 18 months." Besley and Bertha needed one another's support more than ever. "She needed his quiet, confident persistence to help her through that period of financial stress and future uncertainty. He needed her adaptability and gameness in what was to follow."[26]

In addition to the formal schoolwork, Besley used his vacations—if one could call them that—to gain valuable field experience. In the summer of 1903, for example, Besley joined a field party of the U.S. Board of Forestry near Pike's Peak, Colorado, and one Christmas "break" found him studying lumbering in northern Maine. Besley's grueling schedule meant that the burden of child-rearing fell squarely on Bertha's shoulders. In the end, the investment paid off. Thanks to "good health, hard work, and encouragement from my wife," Besley completed his master's degree, graduating cum laude in June 1904.[27] Better yet, his skills were in high demand: Pinchot was hiring every forest school graduate who could pass the U.S. Civil Service examination. Having cleared this hurdle in March 1904 Besley, with degree in hand, was back in the employ of the federal government.

There was little time to celebrate, let alone rest. His entry in the October 1913 issue of the *Yale Forest School News* describes his first assignment: "After graduating from the Yale Forest School in 1904 I was assigned to the Government planting station at Halsey, Nebraska, where I spent one long year of ten months, being in charge of the nursery eight months of the time, during which we planted in the sand hills about 350,000 pine trees." Unfortunately, his efforts did not pay off: "I have since learned that most of the trees died or were burned up by prairie fires some years later. No doubt the large percentage of loss was due to the necessity of using one-year-old stock, since we had no other and had to plant, rather than to the system of planting."[28] After Nebraska, Besley was transferred to the Pike's Peak region of Colorado in March 1905. His job there was "to establish forest tree nurseries and to carry on experimental planting operations."[29]

Meanwhile, important developments were taking place in Washington, DC. In 1905, management of the nation's forest reserves was transferred from the Department of the Interior to the newly created Forest Service within the Department of Agriculture. Now under Pinchot's control, the old forest reserves were organized into national forests. As Besley soon learned firsthand, many of the changes that ensued were unpopular with the general public: "The cattle and sheep men, who had enjoyed free grazing privileges, were particularly incensed when grazing on the forests was put under regulation and fees charged for their use."[30] When his summer forestry work wrapped up, Besley relocated to headquarters in Colorado Springs, where he served as "the local representative of the Forest Service in extension and educational work." It was here, at a meeting near Glenwood Springs, that he was "advised not to appear on the program as it might precipitate trouble."[31] His first brush with an unsympathetic audience left a lasting impression. In the future, Besley would always be mindful of the special problems that forestry professionals confronted when it came to interacting with the general public and, even more importantly, to devising and promoting strategies for conserving timber on privately held lands.

Later in the year, Besley was invited to deliver a presentation at the Annual Meeting of the Colorado Forestry Association in Denver. Unbeknownst to him, Pinchot was among those in attendance. They chatted briefly at the conclusion of the program and then parted company. Besley thought nothing more of the brief encounter. Then in May 1906, Besley found himself working out of a tent camp in a remote section of Pike's Peak National Forest, ten miles from the nearest telegraph station. One day, a telegram for Besley was delivered by horseback. He was offered the position of State Forester of Maryland. In an interview conducted in 1956, Besley maintained "that he never did know why he was selected for the Maryland job." Moreover, he was "curious to know how they found me, planting trees in Clementine Gulch at 10,000 feet in Colorado."[32]

In a memoir penned the same year, however, he wrote: "The offer came from Mr. Pinchot, Chief Forester, who had been asked to recommend a qualified man and who was guaranteeing part of the salary. It appears I was selected because I had taken my academic work at the Maryland State College and because of my good record at the Yale Forest School and later field work in Nebraska and Colorado known to Mr. Pinchot."[33] The latter account is supported by a newspaper report from 1960 that states: "When the Forest Board and the Governor began looking for a director, Gifford Pinchot, the Pennsylvanian who led the fight for conservation in the early 1900's told them: 'The young man you want is Besley.'"[34]

At first Besley was cautious. "[The offer] came as a complete surprise, as I knew nothing of the situation in Maryland or that such a position was even being considered. I replied that I was much interested but wanted to know more of the particulars, especially if the position offered freedome [sic] of action

Fig. 2.3. Fred Besley and his wife, Bertha, ca. 1930, with three of their children: (left to right) Jean, Helen, and Kirkland. Their fourth child, Lowell, is not pictured.

from political control." Although self-identified as an "independent Democrat," Besley wanted to ensure that the fortunes of his department would not rise and fall depending on who occupied the governor's mansion or which party controlled the legislature in Annapolis. Though not a native Marylander, he was familiar enough with the state's rough and tumble political reputation to proceed with caution. Reassured, he accepted the appointment from Governor Edwin Warfield at a salary of $1,500 per year, of which $300 was paid by the U.S. Forest Service "as collaborator." Besley was now Maryland's first, and only the nation's third, state forester. He and Bertha packed up their belongings once again and with two children in tow moved to Roland Park in Baltimore. The peripatetic life of a student assistant and later as an employee with the U.S. Forest Service was now behind them. And, once again, Gifford Pinchot had played a prominent role in shaping the career plans of the former grade school teacher from Virginia.[35]

"Real Pioneering"

Thanks to the Maryland Forestry Conservation Act of 1906, Besley's duties as the state forester were clearly spelled out. Under the general supervision of the newly established State Board of Forestry, which consisted of seven members—the governor, the comptroller, the presidents of Johns Hopkins University and Maryland State Agricultural College, the state geologist, a citizen "known to be interested in the advancement of forestry," and a "practical lumberman engaged in the manufacture of lumber within this State"—the state forester assumed responsibility for "all forest interests and all matters pertaining to forestry and the forest reserves within the jurisdiction of the State." More specifically, this meant enforcing all laws relevant to forests and woodlands, collecting data on forest conditions, directing the work of forest wardens, taking any necessary action permissible by law "to prevent and extinguish forest fires, " directing the "protection and improvement of State parks and forest reserves," and cooperating with various entities, both public and private, "in preparing plans for the protection, management, and replacement of trees, woodlots, and timber tracts. . . ."[36] The new state forester was also required to present a series of lectures on the topic of forestry at the Maryland State Agricultural College.

Although his responsibilities were clearly delineated, Besley knew that putting these directives into action would be easier said than done. There was no existing foundation on which to build the forestry program. There were no precedents or guidelines to serve as models. A miniscule budget ($3,500 per annum to start with) placed severe limitations on what could be accomplished, at least at first. As Besley later put it, it was "real pioneering" work that was carried out in these early days.[37]

Yet there was reason to be optimistic. After six years on the job, Besley put the modest advances of Maryland's fledgling forestry program into perspective: "The progress of forestry has been relatively slow in the southern states and compared with that in the north it may seem exceedingly tardy. In Maryland, while progress has not been rapid, it has been of substantial growth, and today the state has a surprisingly detailed amount of information concerning its forests, its minerals, its agricultural possibilities, and its water resources. Plans for the conservation of these natural resources have been carefully worked out." Besley went on to say that one of the biggest challenges he faced was convincing people of the merits of sound forest management: "I have found as probably many other state foresters have found that while it is comparatively easy to see the line along which developments should proceed, the difficulty is to get the people who should be most interested and who could do the most good to themselves and others, to follow along the lines indicated by good public policy." If there was one very positive development Besley could point to in 1913, it was that appropriations from the state were increasing. In 1913, the annual appropriation for the department was "$10,000 for general purposes, with special appropriations including the purchase of land, amounting to $64,500, for the last two-year period." Compared to other states situated "south of the Mason and Dixon Line," Besley had good cause to "feel satisfied with the results secured in Maryland."[38]

Even with funding and technical support in short supply back in 1906, Besley pronounced his new set-up "very satisfactory." He was particularly impressed by the membership of the Board of Forestry, most notably State Geologist William Bullock Clark, who served as its executive director and as primary contact with the state forester. A few years into the job, Besley remained committed to the arrangement, thanks in large part to the independence he was given: "Under this board I have been given a free hand in developing the forest policy of the state, so that all of the mistakes that have been made are those for which I am personally accountable." Besley's affinity for Clark likely stemmed from their shared interest in forestry issues. As state geologist, Clark had compiled comprehensive reports on Maryland's natural resources including forests. Though they lacked detail and described forest conditions in just a handful of counties, these studies "furnished a kind of starting point for the forestry work." Reasoning "there could be no intelligent, sound program of forestry" without comprehensive baseline information "concerning the character and extent of the forests of the State," Besley decided that his first major undertaking should be a detailed, county-by-county survey of Maryland's forest resources.[39]

Considering the modest resources at his disposal, it was an ambitious project to say the least, but Besley was undeterred. Using U.S. Geological Survey topographic sheets "showing all roads, houses and places" as base maps, Besley, "with the aid of a few forest students available for the summer months," crisscrossed the state by train and via horse and buggy, carefully recording the location, condition, and characteristics of every woodlot five acres or more in size. He then produced final maps at a scale of one mile to the inch. Stands of hardwood were depicted in red and divided into three broad classes: merchantable, sapling, and culled. Pine stands were displayed in green and broken out by species and size. Hardwood-pine mixtures were shown using a combination of red and green.[40] In total, the surveying project took six years to complete.

Along the way, Besley and his team collected "a vast amount of useful information" from "woodsmen, informed citizens, county officials and others."[41] Although one can only speculate, it is altogether possible that Besley had more in mind than mere "information gathering" when it came to these initial contacts with the general public. With the Colorado experience still fresh in his mind, he may have reasoned that the time spent in the field in support of the surveying effort would afford him an opportunity to establish a cooperative— and friendly—relationship with private landowners, local government officials, and members of the press. After all, good public relations was a key ingredient of any successful forestry program, as Gifford Pinchot acknowledged in the fourth edition of *The Training of a Forester* in 1937: "In a peculiar sense the Forester depends upon public opinion and public support for the means of carrying on his work, and for its final success."[42] Besley may have also been aware of the fact that he would need to recruit forest wardens from the ranks of small communities around the state, and this extended time in the field would give him a chance to begin to identify and screen candidates.

Even before the survey was complete, Besley boasted, "Just pride may be taken in the fact that Maryland has more detailed and accurate information concerning her forests than is known concerning the forests of any other State."[43] Later, he would maintain that "Maryland was the first state to compile and publish a detailed, accurate map and report on a state's forest resources."[44] At first, Besley hoped to publish special reports and large-scale color maps for each of Maryland's twenty-three counties. Although several highly detailed maps were published, the scheme soon proved too costly to carry out as originally planned. Besley, therefore, reduced the scale of the maps, condensed the county reports, and assembled them into one volume entitled *The Forests of Maryland*, which was published in 1916. Expanded and updated versions of these county reports were later circulated, often in cooperation with the Maryland Geological Survey, along with numerous other technical reports. Taken together, these documents provide invaluable information regarding the condition of Maryland's forests during the first two decades of the twentieth century as well as the driving forces responsible for deforestation.[45]

These reports and maps revealed that Maryland's forest cover had been greatly reduced since Euro-American settlement and that the remaining forest resources of the state were in relatively poor condition. Whereas forests once covered "at least 90%" of Maryland's land area, by the time of the survey they occupied just thirty-five percent and much of this was "brush land, bearing no merchantable timber of value." The reports also indicated that the "demand upon the forest capital" greatly exceeded the supply of timber and, further, that the state's cut-over forests were in such a deplorable condition and so badly managed that their future productivity was at risk.[46]

In 1910, Besley placed much of the blame for this sad state of affairs on rampant forest fires, contending that they were ". . . accountable, in a large measure, for the poor quality of forest produce and the low yields, by checking the growth, and causing defective trees."[47] A significant effect of forest fires was that they caused reductions among less fire-resistant species.[48] Besley also concluded that Maryland's highly developed transportation network and proximity to key markets—including New York City, Philadelphia, Richmond, Washington, and Wilmington—accelerated the process of deforestation.[49] It was also evident that human activities were responsible for altering the distribution patterns of individual tree species. In southern and eastern Maryland, pine invaded abandoned fields, where in former times the trees had not existed. Out west, white pine had been all but eliminated as an important commercial tree.

From a geographic perspective, southern Maryland was deemed to have undergone the greatest transformation "because it was in this section that the first settlements were made." In St. Mary's County, for example, the "original forests" had "suffered so much from unconservative cutting that . . . only the less important timber species are represented on large areas." Across central Maryland, clearing was more gradual. This section of the state, however, possessed the smallest percentage of forest cover. Besley's remarks concerning the forests of Queen Anne's County are representative: "Of the original forest, very little is left—less than one per cent—and this consists of a few tracts of considerable size that have been preserved largely through sentimental reasons. Practically all of them have been cut over—some of them as many as four times in the last fifty years." At the time of the survey, western Maryland maintained the highest percentage of forest cover due in large part to the area's unsuitability for agriculture and difficult mountain terrain. Yet, even here, steam-powered sawmills were penetrating deeper and deeper into the woods. Westernmost Garrett County, according to Besley in 1914, "at one time contained a magnificent virgin forest of white pine, hemlock and mixed hardwoods, principally oak, maple and chestnut. . . . Today there are only a few small areas that remain in the virgin condition."[50] With this valuable baseline information in hand, Besley's next challenge was to tackle the forest fire problem and stem the tide of indiscriminate and inefficient timber harvesting.

Forest Fires—A "Serious Menace"

During the first two decades of the twentieth century, unchecked fires posed perhaps the greatest threat to the state's—and nation's—forests. If coverage in the pages of the *Baltimore Sun* and *Baltimore News* is any indication, concern over the damage was not limited to forestry professionals.[51] How critical was the problem? In Maryland, Besley emphasized fire as a "serious menace" in many parts of the state. Assessing western Maryland's forests in 1900, State Geologist Clark commented that the "prevalence of fires, following the severe lumbering, has greatly deteriorated the quality of the reproduction and second growth, so that the outlook for a valuable future crop is, at present, not bright."[52] Appraising the condition of Anne Arundel County's forests in 1917, Besley pointed to fire as the "chief source of damage to the forest," noting that there was "a general lack of appreciation of the damage that fires do. People are frequently careless in this matter. In consequence most fires are the result of carelessness, and as the damage is not fully appreciated, the actual conditions must be forcibly expressed and the education of public sentiment encouraged fully."[53]

To combat forest fires, Besley attacked the problem from all sides. He waged an all-out campaign to educate citizens, he lobbied successfully for the

Fig. 2.4. "Forest Destruction from logging Keyser's Ridge, Garrett Co.," March 1929. The introduction of mobile cutting equipment, such as steam-powered circular saws and band saws, meant that even remote sections of forest could be exploited by logging companies.

Fig. 2.5. Planting on lands of F. F. Nicola after clearcutting, ca. 1920s. Destructive fires often followed in the wake of logging. Sometimes "brush burners" or careless farmers and cattlemen were responsible for the blazes.

Fig. 2.6. This new fire tower, ca. 1920s, was constructed in metal. Fire towers were placed strategically across the state. If a fire could be detected early, its damage could be limited.

construction of fire towers, and hired hundreds of forest wardens. These local community members "served without salary" but received hourly wages for the actual time they spent fighting fires. The same is true of the men they enlisted to assist them. According to Besley, ". . . a large number of forest wardens distributed in strategic locations throughout the state was needed to take prompt measures to suppress forest fires which were especially prevalent in the spring and fall in every section."[54] These wardens served a dual purpose in Besley's mind. Not only were they the first line of defense against forest fires, but they also served as the face of the forestry department in the local community.

Besley was careful to avoid politics when it came to the management and day-to-day operation of his department. In 1956, he boasted that, during his thirty-six years in office, "not a single employee of the Department owed his appointment to political influence. Each was a personal selection including some 1200 Forest Wardens, scores of clerks and technical personnel. It was a grave responsibility to assume but has resulted in highly qualified, loyal personnel to which the success of the forestry work is mainly due." "Politics" was an especially sensitive subject when it came to getting forest warden commissions approved. Assured that "there was no political significance to the office," commissions were generally issued in a timely fashion, although delays occurred initially with each new administration.[55]

Besley maintained that forest wardens under his direction were indispensable when it came to fighting forest fires. His entry in the October 1915 issue of *Yale Forest School News* credits "certain state fire wardens who are Dunkard preachers and men of great influence in the community" for progress in fighting forest fires in western Maryland.[56] Others were more skeptical of the work performed by the forest wardens. On one memorable occasion, the actions of forest wardens serving in Frederick County were called into question. Claiming that fire frequency had actually increased since the forest warden system had been organized—a charge that carried with it the suggestion that the wardens might be setting fires so they could get paid to put them out—the Frederick County commissioners and their allies introduced a bill in the state legislature that would exempt Frederick County from having to pay the cost of fighting forest fires. In a confrontational hearing before the Judicial Proceedings Committee of the Senate, Besley defended forest protection and the warden system; he also claimed that these "self sacrificing" men were protecting property and preserving forest resources and the apparent increase in fires could be explained by the fact that, before the wardens were hired, "fires went unchecked and unnoticed," whereas later "every fire received prompt attention and was brought to notice." The climax came when Besley displayed the list of forest wardens for the county: "As I called each individual name and asked if he was not the type of man who could be trusted, not a single protest was made—all were above reproach." In the end, the bill was dropped.[57] Although the county commissioners lost their case, the aforementioned law requiring counties to absorb the cost of fighting fires was soon amended so that the state bore half of the burden. Later, passage of the Weeks Act in 1911 and the Clarke-McNary Act in 1924 made federal funds available for fire fighting.

A similar story involved a grievance that reached the desk of Governor Albert C. Ritchie: "A complaint came . . . from a democrat in Garrett County seeking a public office that Forest Warden A. L. Sines, one of our oldest and most reliable men who had a special assignment on one of the State Forests, was partial to his republican friends in employing help in fire control work. I sent excerpts from the letter with a request to A. L. Sines for a report. The report came back something like this (with spelling corrected): 'I do not mix politics with my job. When we have fires and need help I get the best men available and don't consider politics or religion. Fighting fires is hard work and we need the best men in both parties.' This was transmitted to the Governor with a covering letter and he was satisfied."[58]

A meticulous record-keeper, Besley kept track of every fire reported in each of Maryland's twenty-three counties. He typically recorded the location and date of the fire, the number of acres that burned, the estimated damage in dollars, the cost to extinguish the fire, and the suspected cause. In 1909, for example, eighty-three fires in eighteen counties burned approximately 21,217 acres resulting in estimated damages of $72,080. Thirty of the fires were doused at county expense. With respect to cause, thirty-two were ignited by railroad engines, twelve by hunters, and twelve by brush burners. Campers and saw mills were responsible for starting one fire each. One fire was "reported as incendiary." The cause of twenty-four fires was not known. By Besley's reckoning, "nine-tenths of the forest fires" were probably "preventable with reasonable precautions."[59]

Fig. 2.7. About 1,000 acres burned in this surface fire in Beltsville, Maryland, on April 17, 1926.

Fig. 2.8. Forest fire fighters during the early twentieth century were often unpaid volunteers, including African-Americans.

Fig. 2.9. Edmund George Prince, ca. 1920s. He served as the Patapsco Forest Reserve's chief warden for three decades. Wardens were tasked with preventing and detecting forest fires and ensuring the safety of campers and other visitors.

Fig. 2.10. "J. H. Sims [and] Mike Tasker. Veteran Forest Wardens, Garrett Co.," March 1928. State Forester Besley recruited hundreds of wardens in all parts of the state to fight forest fires and maintain roadside trees.

Fig. 2.11. "Standard Fire Tools, Patapsco State Forest," May 14, 1928.

Fig. 2.12. Resident Wardens Conference at Herrington Manor, July 11-16, 1932. First row (left to right): Robert O'Keefe, Abraham Lincoln Sines, and Fred Besley. Second row (left to right): Henry C. Buckingham, David O. Prince, and M. Carlton Lohr. Third row (left to right): Matthew E. Martin and Grover Cleveland Mann.

Fig. 2.13. "Fire Crew on way to fire–Patapsco State Forest–Ilchester, Md.," May 14, 1928.

Thirteen years later, Besley reported 256 fires across the state. Although the number of fires increased dramatically between 1909 and 1922, partially due to more accurate reporting, the total acreage burned, 20,645 acres, was actually smaller in 1922 than in 1909. The total cost of damages incurred was $92,814.38. The story was quite different the following year. In 1923, 455 fires charred 87,553 acres with damages topping out at $288,934.06.[60] Writing in 1928, Beatrice Ward Nelson stressed that fighting forest fires "has been a major activity of the department since its beginning." However, she also noted that, although the "present forest protection system covers the entire state," it is unable "to fully control the problem."[61]

Reforestation and Recovery

In addition to fighting fires, Besley and his staff promoted reforestation efforts. Sometimes their assistance took the form of advice and the provision of trees from the state nursery. On other occasions, they adopted a more formal approach. One such effort centered on the planting and protection of roadside trees. Maryland's Roadside Tree Law, passed in 1914, is purported to be "one of the oldest urban forestry laws in the United States."[62] Essentially, it permitted the State Board of Forestry "to plant trees along the roadsides, to protect roadside trees, [and] to establish one or more nurseries for their propagation." In a further attempt at beautification, the law also sought "to prohibit the unauthorized placing of advertisements and other notices on the public highways or the property of other persons. . . ."[63]

In a contributing article to the *Journal of Arboriculture* in 1978, former Maryland State Forester Tunis J. Lyon remarked that the preservation and protection of Maryland's roadside trees were a matter of public concern in the early twentieth century: "It was because of this public concern and the foresight of Maryland's first State Forester, Mr. Fred Besley, and the strong insistence of garden clubs and influential women's groups, the Maryland Legislature, some 63 years ago, passed a law that recognized the need for control of these roadside trees. It was then known as Chapter 824, Acts of 1914, and is now, with little or no changes since its inception, found in Article 5 of the Natural Resources Code."[64] Ralph Widner asserts that Maryland's Roadside Tree Law "was the most advanced legislation of its kind" at the time. Lyon maintains that it "was the first state-wide legislation in the United States to plant and protect roadside trees and prohibit unauthorized advertising on public highways."[65]

Enforcing the roadside tree regulations proved difficult at first. According to Widner, "no appropriation was made by the General Assembly for its enforcement," an omission that led many to believe that the law would be rendered "inoperative." Strict enforcement was achieved, however, when Besley tapped a small group of forest wardens to receive special training from a "tree expert" so they could qualify as "tree supervisors."[66] Volunteers also supplied assistance. On "Sign Board Day," June 21, 1914, women's organizations turned out "with Carrie Nation fervor" to rid the state's highways of unwanted signs and enforce the new law.[67] Clearly, the Roadside Tree Law struck a popular chord with many members of the general public.

Efforts to protect roadside trees and generally enhance the appearance of Maryland's highways received an additional boost from Governor Ritchie. In a speech carried in the April 8, 1920 edition of the *Oakland Republican*, the governor was effusive in his praise of the program: "What a magnificent thing it will be for the next generation to have our roads lined with branching oaks, elms and other shade trees. Not only will these trees be ornamental, but they will also be factors in the elimination of dust and dirt. With shade trees lining her roads Maryland would be the most artistic state in the Union. As we have the best roads, we have also the beginning of the most artistic vistas of roads. I trust that residents along our public highways will seize Arbor Day as an occasion for the transplanting of saplings from nearby woods on the sides of roads."[68] As Governor Ritchie's speech suggests, Arbor Day was sanctioned as a particularly auspicious occasion to recognize the value of trees and tree planting.

Press coverage of Arbor Day events as well as numerous other forestry initiatives point to the critical role the print media played in advancing the work of the State Board of Forestry. Whether the topic was school children planting trees on Arbor Day, the benefits of the Roadside Tree Law, or the popularity of the state nursery, newspapers offered generally favorable reviews of the work being conducted by the state forester and his assistants.[69] Perhaps with this in mind, Besley and his staff took advantage of every opportunity to engage the general public.

Perusing the reports of the State Board of Forestry and, later, the State Department of Forestry, one gains an appreciation for the time and energy spent on education outreach. In the report for 1924 and 1925, Besley wrote: "A popular service rendered by the Department is the giving of illustrated lectures on various forestry subjects by the State Forester and his three assistants. For this purpose there are available 1,000 lantern slides, most of them colored, and several reels of motion pictures, with proper equipment for showing them."[70] A humorous note from the pages of the April 1923 *Yale Forest School News* spoke to Besley's versatility as an actor: "F. W. Besley and Josh Cope, '14, appear as actors in the recently released film, 'Pines that come back.' Besley takes the part of a farmer who is persuaded to take up forestry on his old fields by Cope, the forester. As reported, there are no love scenes, clinches or other adaptations to the average mind of the movie fan. Probably Besley wouldn't stand for it."[71]

Besley's skills as a photographer, cultivated as a field assistant in Kentucky and sharpened during his tenure in Maryland, were valuable assets to his old forestry campaign. In her 1928 review of state parks, forests, and game preserves, Nelson noted that "the department finds the education of the public in forestry work an important phase of its administration."[72] No venue was too remote, no gathering was too small. Whether it was a meeting of the Rotary Club in Havre de Grace, a troop of Boy or Girl Scouts in downtown Baltimore, the Wicomico Grange in Salisbury, or an assembly of parents and teachers in Timonium, the state forester and his staff were ready and willing to champion the cause of conservative forest management. Newspapers again played a prominent role by advertising lectures, covering events, and, on occasion, offering unabashed endorsements. The following editorial published in the 18 March 1921 edition of the Bel Air paper illustrates the support of the press: "At the request of THE BEL AIR TIMES, one of the most expert foresters in the State was requested to give his views on forestry in Harford, and the following article is the result. We hope that farmers will read it and adopt plans for replenishing our rapidly disappearing timber supply."[73]

Besley's reforestation campaign manifested itself in other ways. Viewing forest management as "the science of making woodlands pay," Besley searched early on for opportunities to demonstrate the benefits of conservative forest management to the private woodlot owner. He developed a program whereby State Board of Forestry personnel brought timber growers into direct contact with timber buyers. The July 1915 issue of the *Yale Forest School News* explained the program succinctly: "In a talk before the Society of American Foresters on May 6, F. W. Besley gave an interesting description of the assistance which the State of Maryland is now rendering to the woodlot owners. Instead of the old-time, cut-and-dried report the man who examines the woodlot now makes an estimate of the amount and value of the merchantable material present, points out how much of this should be cut, and gives the owner a list of dealers in his locality, and draws up a sample contract for the sale."[74] Several months after the "Timber Marking Plan" was put into effect, Besley proclaimed it a great success: "[I]t has brought such satisfactory results that it is being offered generally as a means of enabling the woodland owner, without experience in handling forestry problems or marketing his forest products, to do so to the greatest financial advantage to himself, and at the same time to secure the improvement of his forest."[75] Nearly ten years later,

Besley pronounced that "interest in forestry" had "shown a substantial gain" due to "the larger number of private land owners who have put forestry in practice on their own land and those who are interested in reforestation."[76] It was a topic he returned to time and again, whether at professional meetings, in the pages of Yale's alumni magazine, or in his official annual and biennial reports.

Forest Reserves and Parks

Another major area where Besley strived to make progress in these early years was in the acquisition of public lands. Section 3 of the Forest Laws of Maryland clearly conveyed the policy of the state on this matter: "[T]he State Board of Forestry shall have the power to purchase lands in the name of the State, suitable for forest culture and reserves, using for such purposes any special appropriation or any surplus money not otherwise appropriated, which may be standing to the credit of the Forest Reserve Fund." The Forest Laws also stipulated "that the Governor of the State is authorized upon the recommendation of said State Board of Forestry to accept gifts of land to the State, the same to be held, protected and administered by the State Board of Forestry as State Forest Reserves, and to be used so as to demonstrate the practical utility of timber culture, water conservation and as a breeding place for game."[77]

While there is little question that the State Board of Forestry was interested in building on the Garrett brothers' bequest, there was virtually no money available for this purpose before 1912. Prior to this time, the only addition to Maryland's public holdings was a small parcel of land on the Patapsco River donated in 1907 by Baltimore attorney John Glenn. Starting in 1912, however, the State Board of Forestry saw its operating budget increase significantly from $4,000 in 1912 to $10,000 in 1913, to $28,580 by 1921.[78] In addition, the Maryland Legislature approved special appropriations to publish Besley's forest surveys, set up a nursery at College Park, acquire Fort Frederick, and begin purchasing properties along the Patapsco River (with the goal of establishing a sizeable forest reserve there).[79]

Recent research suggests that the state board of Forestry's improved fortunes came about as a result of an alliance that combined the scientific forestry concerns of the State Board with the social and recreational interests of influential members of Baltimore's economic establishment.[80] The arrangement bore fruit almost immediately as the state board added twelve properties to the Glenn reserve over the next three years.[81] Besley's entry for the October 1915 issue of the *Yale Forest School News* confirms this: "The state has secured 1000 acres in the new State Park and Forest Reserve along the Patapsco river and is cooperating to protect other areas belonging to the water company and to private owners."[82] These "other areas" were later called auxiliary forests (lands that continued to be owned privately but were managed and protected by the state).

Although confronted with more pressing problems, Besley firmly believed that any self-respecting forestry department needed to own and manage forests—if only for the purpose of demonstration—in order to achieve success. Besley forcefully reiterated this point in the late 1920s when he became embroiled in a debate over whether or not Maryland should repeal the enabling act that permitted the federal government to establish national forests within its borders. Like his old friend William Bullock Clark, Besley also suspected that some lands, if left in private hands, would never fully recover. This was particularly true in western Maryland. Here, Clark opined, "forest management . . . could best be carried on by the State rather than by private owners, as the long rotation required in this section to mature timber would not be as objectionable to the former as to the latter. . . ."[83] Besley was less circumspect: "The land could be purchased at low cost, and under State control and protection it could be made a valuable asset. . . . The purchase of such lands would be an investment and not an expense since they would eventually pay back all costs from the revenue derived."[84] McCulloh Brown also supported forest reserve expansion in the western counties: "Declaring that there is small chance of hardwood timber stands in Western Maryland developing a maturity under private ownership, W. McCulloh Brown, president of the Maryland Forestry Association, last night proposed that the State take over at least 200,000 acres of land not adapted to agriculture in that section for production of timber as a State resource."[85] Thus it was here, in western Maryland that the Garrett donation was followed by one from Henry and Julian LeRoy White, who handed down their Herrington Manor estate in 1917. The state then purchased a fifty-seven-acre block connecting Swallow Falls with Herrington Manor.[86]

The acquisition of these properties begs the question: How were they to be utilized? While historical evidence points to multiple uses, it is patently obvious that Besley's training predisposed him to emphasize the conservation value of the state forest reserves; hence his desire to make them as productive as possible. In this way, the state reserves could serve as shining examples to private woodlot owners who still needed proof of the economic benefits of conservative forest management. Besley's recommendations for Allegany County epitomize this viewpoint: "[I]t is . . . of the greatest importance to the county to make these lands as productive as possible. This is further emphasized by the need of a good local supply of timber to carry on the present industries, and to aid in their further development. The extensive coal mines in the western part of the county require immense quantities of mine props, pit ties and mine rails; the railroads draw upon the forests for large quantities of cross ties; the telephone and telegraph companies require thousands of poles annually; the saw mills and wood-using industries, with large amounts of capital invested and giving employment to hundreds of men, cannot be maintained without a cheap and abundant supply of timber."[87] Besley's position was unambiguous: the state needed a steady supply of timber to support future industrial growth. Whatever the state could not supply on its own would have to be imported. Would it not make good economic sense to make the state's forests—public and private—as productive as possible? To Besley, the answer was obvious.

His fixation with forest productivity notwithstanding, Besley knew that publicly owned lands could satisfy other needs as well, including recreation. In fact, the Forest Laws of 1906 were clear on this point: the state forester was responsible for the "protection and improvement of State parks and forest reserves." This mention of parks would seem to indicate a recreational purpose for the forest reserves. Although Maryland lagged behind other states in the establishment of state parks, State Board of Forestry documents, newspaper reports, and Besley's correspondence with the *Yale Forest School News* show that the forest reserves, or at least portions of them, were touted as parks from their inception.[88]

In *The State Reserves of Maryland: A Playground for the Public*, Assistant Forester J. Gordon Dorrance emphasized that the general public was more than welcome to explore and enjoy the forest reserves. "The term 'Reserve' means, literally, some place kept in store, held back for future use. It is the intention of the Maryland Board of Forestry that it shall practically apply as reserved, but for public use now. It is very well to safeguard the water, and protect the land; but modern forest practice has its best office in making actual contribution to the public weal and wealth. It is with this thought that the State Reserves of Maryland are thrown open for generous use by all the people of the State."[89] Another passage from this document is particularly valuable as it refers three times to the Patapsco Valley State Reserve as a "park":

> Nearer to Baltimore, so near, in fact, as almost to be called a city park, is the Patapsco State Reserve. Maryland owns here 916 acres, chiefly of wooded land, with the addition of over 1,000 acres which are open to the public, with full park privileges in return for the protection which the Board gives to its respective owners in the matter of patrol against trespass and fire. The entire Reserve is essentially a protection and a recreation forest. Prior to 1912 this region was only a piece of attractive country: two high, sloping banks with a cover of timber, a winding river between; it was close to Baltimore; it seemed to have some natural possibilities as a park; and its forests covered and protected the watershed of the Patapsco, thus affecting in a measure the harbor of the city. . . . Under the management of the Board its attractions are being protected and so far as possible enhanced, and the Patapsco Reserve made ready for free use by the people of this State.[90]

Shifting his attention to Garrett County, Dorrance promoted the recreational potential of western Maryland's "five large forest parks":

> That no more people have seen them is not so strange, for they are not part of a thickly peopled district. Recently, however, with the acquisition of the Patapsco State Reserve, the Maryland Board of Forestry, in whose charge these lands are placed, became convinced that if the people of Maryland had a better understanding of how to enjoy the five large forest parks within their reach the knowledge would stand them in good stead when it came to the investing of a vacation which might be spent on any part of several thousand acres offered free for use and readily and cheaply accessible from any point. Located in Garrett County, in the higher altitudes of Western Maryland, the Skipnish,

Fig. 2.14. Swallow Falls State Park, ca. 1958. Although a utilitarian conservationist by training, Besley recognized the potential of state forests to serve a recreational purpose as well.

Kindness, Swallow Falls, and Herrington manor Reserves will appeal to those who like their vacations seasoned with a little wild life, a dash of the woods and the mountains, and withal a vivifying atmosphere.[91]

Again, although these parcels were officially classified as forest reserves, the state dubbed the areas "parks" in promotional literature.

Newspaper accounts from the 1920s also sustain the idea that the State Board of Forestry was interested in recreation. A story carried in the *Evening Sun* supports this contention: "The development of the Forest Reserve along the Patapsco as a recreation and camping ground for Baltimore people, which got well under way

Fig. 2.15. Swallow Falls State Park, ca. 1958. The park's towering hemlocks, enormous rhododendrons, and rugged beauty have attracted tourists for decades.

last year, will be pushed during the present season by State Forester Besley. . . ." Subtitled "Patapsco's Pretty Scenery And Natural Beauty Thrill Campers And Nature Lovers," the article is particularly notable for its romantic imagery and for its reference to Besley's own involvement in outdoor recreational activities:[92]

> There is a healing touch in the contact with the spirit of Nature, and especially at the time of midsummer fullness, quiet and peace. . . . State Forester F.W. Besley and his assistant foresters are preserving this wild, natural beauty. They are making it possible for Marylanders who cannot go as far as Maine or Canada to get as close to Mother Nature as in the wild and unexplored regions of the North. On the slopes rising up from the river in thick virgin forest land,

traversed by springs and streams, ideal camp sites have been staked out. Some derive their beauty from the view; others from proximity to the river; and some because they are built right on the edge of a leaping mountain cascade. . . . Mr. and Mrs. Besley and two of their children are spending a month in a camp overlooking the upper most rocky basin of one of these lovely cascades. . . . To be entertained at the Besley camp is a pleasure long to be remembered. Not 10 feet away from the open-air dining tent the water rushes over the rocks of Upper Falls. One goes to bed in the big Army tent, with its "double-decker" cot in the middle and its pine needle couches on either side, to the sound of this music and wakes up with it in the morning. . . .[93]

The Besleys also camped regularly in western Maryland. Reminiscing in 1996, Helen Besley Overington looked back with fondness on the summertime trips she took with her family in the 1920s. Accompanied by Assistant Forester Karl Pfeiffer and his family, the group often stayed at Herrington Manor: "Mrs. Pfeiffer (assisted by HBO) and Bertha Besley (assisted by daughter Jean) did the cooking for the men. It was 5-6 miles to Oakland, so you had to take the groceries with you. . . . Link Sines sometimes came over at lunch time, and the Besley children ate his lunch and Link ate their lunch. Link came with a fried egg sandwich, which the Besley children had never heard of and thought that was great." Overington remembered that her father "was responsible for having the log cabins built and was very proud of them since he helped design them. They became very popular." She also recalled a time when "Ford, Edison and Firestone took one of their famous camping trips to Swallow Falls. The shining new Fords attracted a great deal of attention among the local populations."[94]

Two questions arise at this point. First, why publicize the forest reserves as parks? Two possible explanations come to mind. It is readily apparent that Besley enjoyed the outdoors and genuinely believed it was a good idea to extend these recreational opportunities to Marylanders. The more pragmatic side of Besley, however, may have calculated that, in order to gain support for his forest conservation initiatives—and for future purchases—he needed to win converts. What better way to do this than to encourage the public to visit the forest reserves, observe the progress that was being made there, and

Fig. 2.16. Camping at Patapsco State Forest, Summer 1921. According to historian Robert Bailey, Baltimore's Hutzler department store reserved camp sites for its male employees and their families.

Fig. 2.17. "Swimming at Vineyard Patapsco State Forest," Summer 1921. Given its proximity to Baltimore, Patapsco State Forest has long served as a recreational outlet for the city's residents.

Fig. 2.18. "Mess Tent Scout Troop 149 Camp site OG16," June 1920. Throughout his career, Besley maintained strong ties to local boy scout groups. Excursions to Patapsco State Forest gave city kids an opportunity to learn about and enjoy the great outdoors.

Fig. 2.19. "Balto Troop 135 at Swallow Falls A. P. King–Scoutmaster," September 1925.

take advantage of the recreational opportunities that were now available to them? The second question is more difficult to answer. Rather than advertise the forest reserves as parks, why not simply establish parks? Even in these early years, Besley may have feared that, by doing so, he would invite calls to establish a new agency with separate funding. Perhaps such an arrangement would lead to competition for land and, more importantly, competition for scarce funds. Such a stance would have been consistent with the one Pinchot was taking at the national level.[95] Also, Besley was used to being in charge. As one relative later put it, he "fought very hard to keep the forest service intact" and "would have resisted any attempt to fragment his authority."[96] He would refer to these lands as parks when doing so suited his needs, but he would, for the time being, maintain their official status as forest reserves. This way, he could be sure that they would remain in his care.

Whether one viewed them as state forests or state parks mattered little. Taken together, the Patapsco Forest Reserve, Fort Frederick State Forest, the state forests of Garrett County (which encompassed the Skipnish, Brier Ridge, Herrington Manor, and Kindness tracts), and the auxiliary forests amounted to less than 5,000 acres. If there was one aspect of Besley's program that failed to meet expectations after more than two decades of professional forestry in Maryland, it was land acquisition. Besley learned the hard way that building a system of public lands can be a painfully slow process.[97]

Growth and Reorganization

Seventeen years after gaining a foothold in the state, professional forestry in Maryland experienced a significant change. In retrospect, 1923 turned out to be a watershed year for Besley's department. Under a reorganization act, Maryland's eighty-five boards and commissions were consolidated into eleven departments. Too small to stand on its own, the Forestry Department now came under the purview of the Board of Regents of the University of Maryland. At the same time, the governor appointed an Advisory Board of Forestry, made up of three members from the former Board of Forestry along with two new appointees. According to Besley, the University of Maryland was selected as the new home for the Forestry Department because "it was non-political and the State Forester

Fig. 2.20. "Forestry tent & Exhibit at U. of M. on Farmer's Day," May 26, 1923. Like his mentor, Gifford Pinchot, Besley took every opportunity to promote the work of the Forestry Department.

Fig. 2.21. "Forestry Exhibit, B & O Centennial," October 1927.

was by law lecturer in forestry in the University."[98] A key provision that ensured a smooth transition was a ruling by the attorney general that the Board of Regents should operate as a body independent of the University of Maryland when acting on matters involving the Forestry Department.

Attempts at reorganization had been made before. An item in the October 1916 issue of the *Yale Forest School News* describes one such effort: "The great progress which this southern state has shown under a consistent policy, free from politics and in charge of a single technical man from its inception, was strikingly vindicated last winter by the utter failure of an attempt by ambitious politicians to absorb this growing and popular department in a so-called reorganization with agricultural interests."[99] The 1923 reorganization differed in two important ways from the failed attempt of 1916. First, it did not appear to have been politically motivated. The second major difference was that changes made in 1923 would affect Besley throughout the rest of his career.

As time passed, Besley became frustrated by the organizational changes. Indeed, the late 1920s and 1930s were especially difficult years for Besley and the Department of Forestry. While professional forestry continued to make important strides, state funding hit new lows, and relations with the president of the University of Maryland deteriorated. To make matters worse, Besley and the Maryland forestry program received stinging criticism from unexpected quarters. Meanwhile, professional forestry was gaining ground in an unexpected place: Baltimore.

Fig. 3.1. Monument Street, Mount Vernon neighborhood in Baltimore, ca. 1904–1915. The desired integration of cultivated nature and sophisticated architecture is self-evident in this park-like urban setting, a legacy that continues in many of Baltimore's neighborhoods today.

Chapter 3

Managing Baltimore's Urban Forest, 1912–1942

City plants must contend with tremendous biological, physical, and chemical stresses: too much water or too little; temperatures too cold or too hot; polluted air, water, and soil; pests and diseases. Many plants cannot survive at all; others survive in a dwarfed, distressed condition. The city contains a remarkable variety of habitats within a stone mosaic of buildings and pavement.

—Anne Whiston Spirn

In 1906, Maryland officially joined the ranks of states establishing professional forestry agencies. The decision to inaugurate such a program was sparked by a private donation of land, made on the condition that adequate means for the protection of the state's forests, both public and private, would be provided. The offer moved a conservation-minded state senator, McCulloh Brown, to draft the Maryland Forestry Conservation Act, which established the State Board of Forestry and the Office of State Forester. By 1923, the Forestry Department enjoyed a favorable national reputation, thanks to State Forester Fred W. Besley, who had built the agency into a model of forest conservation, although there would be bumps in the road ahead for Besley and the program.

Less well known is the city of Baltimore's early experience with professional forestry—a neglected piece of forest management history. Unlike the forests administered by the state, Baltimore's urban forest was not managed with timber production in mind. Rather, trees were planted and maintained primarily for aesthetic purposes. Instead of worrying about fighting forest fires, the city forester and his assistants were concerned with activities such as pruning, spraying, and watering. And, while both the State Board of Forestry and Baltimore's Division of Forestry were engaged in reforestation efforts, the task of growing a tree in an urban environment differed greatly compared to a more rural setting.

Even the origins of the two programs differ markedly. While both the state and city forestry programs trace their roots to the Progressive Era, their paths diverge from there. At the beginning of the twentieth century, statewide forest conservation practices in general—and acquisition and management of public lands in particular—were concepts that had to be "sold" to a largely indifferent general public by a small group of state government officials. In contrast, professional forestry in Baltimore was readily accepted by city residents, many of whom took up the task of "selling" the idea to the city's famously unresponsive municipal government. Civic organizations and neighborhood associations, in particular, played a prominent role in spearheading the drive to establish an agency to care for and manage the city's trees. Civic enthusiasm notwithstanding, by the 1940s, Baltimore's urban forest appeared to be in decline.

Historical Antecedents

Throughout the nineteenth century, but especially after the Civil War, trees were planted along streets and highways and in public parks and squares in and around Baltimore to promote a more bucolic atmosphere. At the time, creating parks and planting shade trees, were viewed as important steps toward countering the dehumanizing effects of the Industrial Revolution. By the early twentieth century, however, it was evident that the needs of the city's trees were not being met and, further, that a more significant commitment of resources was needed to protect them. As contemporary newspaper accounts and annual Board of Park Commissioners reports show, it was during this period that the city of Baltimore, taking its cue from influential members of the general public, first embraced the idea of professional forestry.

One of the first large-scale attempts to plant trees in Baltimore happened in 1827. William Patterson, a successful merchant, "originated, for Baltimore at least, the idea of giving land for urban embellishment and public recreation" when he donated 5.96 acres on Hampstead Hill to the city for development of a "Public Walk." Prior to his death in 1835, Patterson arranged for the planting of

Fig. 3.2. Patterson Park, ca. 1890. Named for Baltimore merchant William Patterson, this tree-lined "public walk" became a city park in 1853.

more than 200 trees "in a dozen varieties." These were arranged in straight rows to provide a "sylvan corridor" for the public walk. In 1853, the public walk was officially dubbed Patterson Park, in honor of its wealthy benefactor.[1] Spurred by Patterson Park's popularity, city officials acquired the "Druid Hill" estate of Lloyd Nicholas Rogers in 1860. Management of the estate's expansive lawns, woodlands, pear orchards, tree-lined drives, and "flock of sheep" now fell to the recently created City Park Commission, which had also assumed responsibility for maintaining Patterson Park.[2]

Perhaps the best source of information pertaining to the management of Baltimore's forested areas during the nineteenth century is the collection of annual reports of the Board of Park Commissioners. While great emphasis is at first placed on Druid Hill Park, later reports detail tree planting and other activities taking place at other parks and along the city's streets. This data set, though incomplete and fragmented, provides us with tantalizing clues regarding the condition of the forested areas in the city's parks as well as the strategies and priorities that park superintendents adopted as they sought to improve Baltimore's urban forest.

The disproportionate amount of space devoted to Druid Hill Park in the Park Commission's reports is, perhaps, understandable when one considers the size and splendor of the property. According to Howard Daniels, the park's first landscape gardener, the city's acquisition of the Rogers estate was a providential event in Baltimore's history. In an 1860 report, he opined: "The preservation of so perfect a natural park, with its primitive growth of noble trees, within the immediate vicinity of a great city, may well be ranked amongst the most fortunate occurrences for the public interest, and in so readily availing themselves of it, the people of Baltimore have added to the noble monuments which adorn their city— one which will gratefully and beautifully commemorate at once their intelligence and their liberality."[3] Daniels' effusive praise was tempered somewhat by the enormity of the project that lay before him. After appraising the health of the new park's trees, he moved quickly to initiate a topographical survey, begin a program of tree planting, and press the case for the establishment of a nursery.

In his *First Annual Report of the Landscape Gardener of Druid Hill Park*, Daniels made the following observation: "The woods and groups now in existence at the Park, are fine, large, and healthy, embracing oaks, hickories, sycamores, tulips and other deciduous trees, fully equalling [sic], if not surpassing, anything to be found in the best public parks of the world. . . ." He nevertheless advocated planting evergreens "judiciously located in masses, mixed sparingly or planted as undergrowth in shrubberies" as a way of "enlivening views of the park during the winter months."[4] Noticing that some of the older trees, particularly those located in the southwest corner, "have been greatly injured by the numerous smaller ones" and that, in several places, "there are three or four times as many trees as can be healthfully sustained in the same space," he advised "thinning out of the less vigorous growth, in order that the more valuable trees may have scope for robust development."[5] With the woods thus thoroughly "cleaned," Daniels recommended seeding the area with grass, ". . . for among the beauties and the convenience of a public park, a clean, even, and luxuriant turf, will be classed second only to the abundant growth of shade trees." Efforts to thin the woods should be concentrated in the southern section of the park, he explained, because this portion of the park was situated nearest to the city. It must, therefore, "receive the most careful attention" and exhibit the "highest cultivated finish."

By contrast, the northern portion of the park "now covered with old and sturdy woods presenting many magnificent forest scenes, should be left untouched in all the grandeur and luxuriance of its primitive growth."[6]

Impressive as the park's "noble trees" were, Daniels was keenly aware that a source of new trees for planting was much needed. To this end, he lobbied vigorously for the creation of a nursery to ensure a steady supply of ornamental trees and shrubs. Noting that very few desirable specimens were available locally, he recommended importing evergreen trees and shrubs from Europe. When funding from the city proved insufficient to set up the nursery, Daniels renewed his request the following year. Reiterating that the supply and variety of trees and shrubs on-site was ". . . far below what will be required for the proper adornment of the Park," he argued that the time was "near at hand when a large number of ornamental trees and shrubs will be required in the progress of improvements already designed." Appealing, perhaps, to a fiscally conservative audience, he added: "The economy of growing trees and shrubs where planting on a large scale is to be continued for years, is a matter of vital importance. Besides the economy of the process, there are also important advantages in having the plants at hand when wanted, and in the greater certainty of their thriving after being transplanted."[7]

Another decision Druid Hill's new landscape gardener faced was what to do with the former estate's 40,000 pear trees. Noting that "their value as ornaments" was "more than over balanced by the irresistible temptation which their fruit furnishes to visitors to violate the Regulations of the Park," Daniels favored a plan whereby the trees would be removed and sold. The proceeds, he reasoned, would "be sufficient to keep the nursery stocked and cultivated."[8] Other sources of income that served to "eke out the means of the Commission" in later years included the sale of cordwood (cut from fallen trees), sheep, and deer.[9] With regard to the latter two, the Park Commission report from 1878 states that the park's "flock of sheep" included two rams and 141 ewes, and that the deer commanded "a ready market in Baltimore and New York."[10] Twelve years later, the Park Commissioners pronounced, "Our flock of Southdowns never was as much in demand as during the past year." According to the 1890 report, twenty-seven bucks and forty-two ewes were sold to butchers for three dollars apiece. Additional animals were sold for fifteen dollars per head.[11]

Although the data are incomplete, evidence suggests that tree planting activities at Druid Hill Park were conducted on a fairly regular basis. Roadside tree planting was viewed as a particularly valuable use of scarce resources, because it enhanced the experience of tourists visiting the park. In 1860, Daniels pointed out that, although much of the park was swathed in "original forest," much reforestation work needed to be done. The following year, he reported that 1,518 trees were planted with "preparations made for many more." The subject of planting came up again in the report for 1865: "Many trees have been planted during the past year in those places in the vicinity of the roads and walks where shade was deemed desirable." And yet "open spaces" and "formerly cultivated fields" remained that required "planting to bring them into keeping with other portions of the domain." Twenty years later, the annual report for 1885 stated that 252 deciduous and 108 evergreen trees were planted at Druid Hill Park. Between 1887 and 1890, the General Superintendent and Engineer of Public Parks oversaw the planting of another 680 deciduous trees and 463 evergreens.[12]

Although trees were planted with the overall goal of reforesting "open spaces" and "formerly cultivated fields," the reports indicate that replacing trees damaged by storms also kept park maintenance crews busy. By 1878, a new threat loomed on the horizon: "A large part of the Park being original forest land, greater or less destruction of trees by heavy storms is naturally to be expected, and the removal of fallen trees, which are converted into fire-wood and sold, gives much work to the small force that the Commission are able to employ. But of late an enemy, worse than the storms, has appeared in the shape of an insect that attacks the hickory tree especially, if not exclusively. . . . During the past year not less than 150 hickories have been killed in this way."[13]

While the lion's share of attention and resources was lavished on Druid Hill Park, modest improvements were also being carried out at Patterson and Riverside parks. In the spring and fall of 1877, for example, 351 trees and shrubs were planted at Patterson Park, while the total number of "shade trees, evergreens and shrubs" planted at Riverside Park "amounted to two hundred and eighty-five, besides some three hundred others in the Park nursery." The following year at Riverside, "a force of twenty-five men and four carts was organized and set to work" planting a total of 427 trees of different varieties, including 216 deciduous trees and sixty-one evergreens. In 1879, 100 trees (deciduous and evergreen) were planted at Patterson Park, along with 182 deciduous and fifty-three evergreen trees at Riverside. Tree planting activities continued in 1880, with 212 planted at Riverside Park. In 1885, seventy-four trees were planted at Riverside, while at Patterson, "The row of sugar maple trees on Eastern avenue was extended the full length of the park, also the row on Baltimore street, both of which were enclosed in painted wooden boxes, and thirty-five (35) of the same variety of trees were planted in different portions of the park."[14]

As the park board reports plainly show, efforts to improve conditions at the city's parks, including expanding their forest cover, were made throughout the second half of the nineteenth century. In the opinion of Frederick Law Olmsted, Jr., these attempts fell short of the mark. In his firm's *Report Upon the Development of Public Grounds for Greater Baltimore*, published in 1904, the renowned landscape architect wrote: "We were not asked to report upon the existing parks of Baltimore, and we have made no investigation of them beyond what was necessary to an intelligent estimate of the remaining needs which they cannot be expected to supply. . . . [O]n the other hand, in view of the foregoing discussion of the purposes which should control in the management of parks as well as their first selection, we feel it our duty to say that in many instances the design and management of the existing parks does not appear to have been controlled by a consistent and far-sighted policy, and that the city has consequently not obtained for its expenditures such large returns in usefulness as might have been possible." In a report issued the same year, the Board of Park Commissioners decried the current condition of the city's trees, particularly those lining its streets. Noting they were "in a shameful condition," the members of the board vowed "to atone for this neglect," though they cautioned that "it is impossible now to entirely correct it, by careful attention this year."[15] What accounts for the deteriorating condition of Baltimore's trees—especially those lining the city streets—at this time, and what solution did city officials and citizens finally agree upon to remedy the situation?

Fig. 3.3. "Mansion House, Druid Hill Park, Baltimore, Md." This postcard depicts the Druid Hill estate of Lloyd Nicholas Rogers. The city of Baltimore acquired the property in 1860 for use as a park.

Fig. 3.4. Garrett Bridge, Druid Hill Park, ca. 1890. Landscapes such as this were inspired by nature, and they became the embodiment of "rus in urbe" (country in the city) ideals that informed urban park planning (in Europe and the Americas) during the Romantic Era.

Fig. 3.5. Busy commercial thoroughfares, such as the East 200/300 block of Baltimore Street, ca. 1870, offered little space for street trees.

Fig. 3.6. "Baltimore street scene." In some residential areas, street tree planting was not a high priority.

Fig. 3.7. The west 200 block of Biddle Street, ca. 1920. For decades, the city's African-American population was confined to a relatively small section of northwestern Baltimore. While the urban forest flourished in neighborhoods such as Mount Royal, they were virtually non-existent on streets just a few blocks away.

Fig. 3.8. Biddle Alley, ca. 1911.

Fig. 3.9. The north 400 block of Calhoun Street. While new housing developments in Baltimore's rapidly expanding inner suburbs assigned a high value to street trees, new construction closer to the city center often neglected to make accommodations for such amenities.

Fig. 3.10. Eastern Avenue looking west from 15th Street, September 17, 1915.

Civic Organizations and Neighborhood Associations Lead the Way

The most obvious explanation for the generally poor condition of the city's street trees is that no department or agency was officially charged with their care. Although the Park Board might have been a likely candidate, the following excerpt from the report of 1904 suggests that they were prevented from doing so: "The Law Department has denied the right to this Board to plant trees along the streets of the city outside of the parks and squares, and probably would also say that this Department had no authority to trim, care for, or in any way expend on these street trees its park funds. Unquestionably, however, a great improvement in them, and consequently in the city's appearance, might be had by subjecting them to some intelligent control." Arguing that Park Department personnel were more likely to possess the skills needed to carry out this task than any other city department, Park Board officials reasoned that such work could be accomplished—and the objections of the Law Department removed—if the city government would "appropriate from the general tax levy a sum sufficient to cover the cost of properly caring for the street trees and place the same to the credit of the Park Department, in addition to the usual park funds."[16]

Under the administration of Mayor Thomas Hayes, who served from 1899–1903, the problem was apparently exacerbated when the city adopted a controversial rule that, according to one newspaper account, compelled "every citizen who desired to plant a shade tree before his residence to come to the City Hall and, like the citizens who desired to keep a dog, purchase a permit or license authorizing him to make this humble effort to beautify that portion of the city in which he was most interested."[17] The rationale for this rule was that any attempt to plant a tree involved tearing up a piece of pavement, and the fee was thus intended to defray the cost of the inspection.

Writing for the *Baltimore Sun* in 1912, the Rev. D. H. Steffens, in an article entitled "What May Be Done Toward Reforesting Baltimore," recalled: "This rule, which took the perfectly tenable ground that a municipality should exercise some control over the trees planted in its streets, raised such a storm of protest on the part of the citizens and was so much ridiculed by the public press that no attempt was ever made to enforce it. Indeed, it is difficult to understand how an otherwise intelligent administration could have attempted to establish and enforce such a rule, which looked at tree planting merely as it affected pavements and the effect of which would have been to penalize rather than encourage tree planting on our city streets."[18]

According to Steffens: "The unanimous refusal of the general citizen body to countenance this penalization of tree planting—for that is what the tax amounted to—is an indication of the sturdy good sense of the common people, their wholesome love of nature and their unerring instinct which associates trees, not only with civic beauty, but with civic health." He then reminded the *Sun's* readers of the value of urban trees: "They purify the air by absorbing the carbonic acid gas exhaled by man and the animals and give back to the teeming life of our cities the oxygen without which that life cannot exist. . . . Trees help to modify the heat of our cities, which is intensified by rows of brick fronts, concrete sidewalks and asphalt pavements, utterly precluding evaporation of water from the earth unless it be through the foliage of the shade trees." He went on: "All real estate men who develop a new section of the city know that nothing is more attractive to a prospective purchaser than a row of shade trees in front of his lots. They, therefore, make every effort to secure a quick growth of trees which are set out as soon as the streets are opened. Trees also have an aesthetic and a moral value. They relieve the harsh, raw outlines of monotonous rows of houses, so tiresome to the eye. They possess an unceasing interest, which varies with every season of the year. They help to give the city dweller that sense of home and attachment to a certain spot of earth out of which local pride, civic virtue and patriotism are born."[19]

A solution seemed possible in 1907 when a series of resolutions recommending the appointment of a commission dedicated to citywide oversight of street trees was drawn up and submitted to Mayor J. Barry Mahool (who served from 1907–1911). According to a *Baltimore Sun* report: "There is reason to believe that in the near future Mayor Mahool will appoint a commission to regulate the care of trees on sidewalks and in the public parks and squares. The suggestion that such a commission be appointed was made at a recent meeting of the Citizens' Improvement Association of Northeast Baltimore. . . . Mr. P. J. Skirvan, recording secretary of the association, said yesterday that it was imperative that a commission be appointed to take care of the trees growing on the thoroughfares of the city. He said that while some persons pride themselves on growing healthy trees in front of their homes, and yearly have them clipped, there are many others who disregard the growth of their trees in every sense of the word." Skirvan then delivered his pitch: "Were a forestry commission appointed, he said, they could make regular inspection of the trees throughout the city and instruct residents regarding their care. This would help beautify the city and would rid it of many dying trees, which are an eyesore to many otherwise desirable neighborhoods."[20]

Mayor Mahool's response was noncommittal: "Your communication of June 18, containing a resolution relative to trees on sidewalks, was received. There is something in the idea recommended by your association that appeals to me. However, while your suggestion contains some value, it occurs to me that there may be some other such schemes which might be even preferable."[21] In the end, the resolution to appoint a commission on trees was dropped.

Public interest in tree planting intensified in 1912 when the local chapter of the Women's Civic League teamed with another neighborhood improvement association to lobby for the formation of an agency devoted specifically to the planting and management of urban trees. This time urban forestry advocates were successful. According to the *Sun*: "After listening to a delegation from the Peabody Heights Improvement Association at the City Hall yesterday afternoon, Mayor Preston announced that he would get behind a bill authorizing the appointment of a city forester, and giving the municipality control over the planting, care and removal of shade trees on streets within the city limits. . . . Data on the subject collected by the Peabody Association during the last 18 months was presented to the Mayor, in addition to the rough draft of a bill. It was shown that a number of large cities had assumed control over shade trees, with good results."[22]

With the backing of Mayor James H. Preston, the city passed the first street tree law, known as Ordinance No. 154, on August 17, 1912. It authorized the city engineer "to regulate the planting, protection, removing and controlling of all trees, growing, planted and to be planted in the streets of Baltimore, not under the jurisdiction of the Board of Park Commissioners, and to appoint a City Forester and such other employees and assistants as may be necessary to carry out the provisions of this ordinance." Additionally, the city engineer was vested with the authority to ensure that all statutes and ordinances governing the "protection of trees in the streets" be "strictly observed." Sec. 3 of the ordinance outlines the responsibilities of private citizens: "And be it further ordained that no person shall plant any tree in any street without first having obtained a written permit therefor [sic] from the City Engineer setting forth the conditions under which such trees may be planted, including the kind and variety thereof, and until the City Forester has designated on the ground the location thereof, and without in all respects complying with the conditions of such permit."[23]

Soon after passage of the ordinance, the Baltimore City Forestry Division was established. The next step was to hire "a person of knowledge and experience in the care and culture of trees."[24] Under the headline "RUSH FOR NEW CITY JOB," the *Baltimore Sun* reported, "Applications for the position of city forester . . . are pouring in upon City Engineer McCay, who is to make the appointment." Although McCay had already heard from a dozen or so men "anxious for the place," he did not expect to name anyone to the position before the first of the year "because I have no money for his salary."[25] In the meantime, city officials went about the business of locating a source of trees for the city: "Returning to the city yesterday afternoon from an inspection of the leading nurseries in Maryland, Pennsylvania, New Jersey and New York, Major Joseph W. Shirley, chief engineer of the Topographical Survey Commission, said Baltimore could buy a fine lot of trees for the sidewalks to be parked and that the cost would be reasonable." Plans were already taking shape to "park" trees on several streets in Northeast Baltimore, North Baltimore and the Annex neighborhoods.[26] That fall, Roger Brooke Maxwell, recently of the Yale School of Forestry, was appointed the city's first forester.[27]

What carried the day for tree advocates in 1912? As the foregoing newspaper excerpts indicate, civic groups, such as the Municipal Art Society, Women's Civic League, and neighborhood improvement associations, played crucial roles. Taken together, these organizations wielded a great deal of political power, especially in the days before comprehensive city zoning. Started in 1899 by several "upper class Baltimoreans," the Municipal Art Society had already achieved notoriety for leading the fight to provide the city with a modern sewer system and for commissioning the 1904 report from Olmsted Brothers. According to Warren Wilmer Brown, "The Society also took an active interest in the conservation of the city's—and region's—trees." Brown, who penned an early history of the Society in 1930, stated that the group "all along has stood for the preservation of trees and other factors of natural beauty."[28]

Groups representing the interests of specific neighborhoods as well as the "greater good" also figured prominently in the passage of Ordinance 154. As Baltimore grew in size and population, neighborhood and community associations—operating independently but also cooperating via a citywide congress of neighborhoods—formed to pressure city officials into dealing with a wide array of problems and to furnish residents with the infrastructure and services they needed.[29] For example, in his recent history of Northeast Baltimore, Eric Holcomb writes that the Northeast Community Association "fought for the extension of the Baltimore County Electric and Water lines to service Hamilton." Not far away, the Arcadia Improvement Association pushed for physical improvements such as extension of sewer lines to the area, removal of trolley poles, installation of traffic and street lights, planting of street trees, and improvements to Herring Run Park. During the first three decades of the twentieth century, the Hamilton Improvement Association, also located in Northeast Baltimore, campaigned to introduce telephone service, gas lines, sewers, and electric lights, eliminate toll gates, improve local roads, and secure more reliable mail delivery.[30] Across town, the Garrison Boulevard Association concentrated on a range of issues such as improving lighting for Garrison Avenue, opening Gwynns Falls Parkway, placing traffic caution signs at dangerous intersections, and extending street car service. They also pressed for improvements to Hanlon Park, the trimming of trees, and representation on the Park Board.[31] By 1910, more than seventy such neighborhood organizations existed in Baltimore.

One group that appeared to exercise considerable influence in city matters was the Peabody Heights Improvement Association (the "Association"). Article II, Section I of the group's constitution identifies the "object" of the Association as follows: "To promote the general welfare of that section of the city bounded by the centre of the following streets: on the South by Twenty-fifth street; on the East by Guilford Avenue; on the North by the City Limits, and on the West by Maryland Avenue." More specifically, the group sought to "secure a compliance with and to prevent a violation of any of the restrictions applicable to that portion of said section known as the Peabody Heights Tract." These restrictions included "requirements that buildings shall be set back 20' from the building line and that no lager beer saloons or places for sale of intoxicating drinks, slaughter houses, bone or glue factories of any kind or nuisances of any description shall be permitted. . . ."[32] The group was also interested in neighborhood and citywide beautification.

According to the meeting minutes of the Association, which span from 1909 to 1933, interest in "planting and caring for trees" dates back to at least March 1910, if not earlier. On May 9 of that year, the existence of a forestry committee is first mentioned.[33] Reference to the work of the "shade tree committee" shows up for the first time in the minutes of the October 10 meeting: "The shade tree Committee made a very lengthy and comprehensive report. . . . The following motion was then carried—That the report be forwarded to the Mayor's office, with the request that the same be sent to the City Solicitor with the request that an Ordinance be drafted in accord with the substance matter of the report; plus the recommendations of the Park Board; and that the Peabody Heights Improvement Association push a campaign, based upon such an ordinance."[34] On February 13, the group resolved that "an Ordinance should be introduced looking to a law governing the planting of trees &c."[35] Almost one year later to the day, the Association threw its support behind the tree ordinance now making its way through the state legislature.[36] On August 17, 1912, Mayor Preston gave final approval to Ordinance No. 154.

Over the course of the next twenty years, the Association enthusiastically supported the work of the Division of Forestry. Embedded throughout the group's meeting minutes are requests for advice addressed to the city forester as well as reminders to send thank you notes to the Division of Forestry for trimming and planting trees in the neighborhood.[37] Of particular interest are the meeting minute notes from November 11, 1912 and December 8, 1913. On November 11, 1912, State Forester Fred Besley visited the meeting and gave "a talk on the shade tree, after which he was given a rising vote of thanks." Thirteen months later, "Mr. Maxwell the City Forester gave an interesting illustrated talk on trees after which a motion was made and carried thanking him for his efforts and also for the planting of trees about the open cut on 26th Street, and assuring him of our support."[38] How well acquainted these two former Yale students were with one another is difficult to say. Any record of correspondence between the two men was either destroyed in the fire that gutted Besley's office in 1920 or has been otherwise lost.[39]

Further evidence of the Peabody Heights Improvement Association's commitment to city beautification is exemplified by its willingness to "endorse the action of the Women's Civic League in regards to their clean up, paint up campaign" and "co-operate with them whenever possible" and to send a representative to the convention of the American Civic Association, a well-known national conservation organization.[40] The activities of the Association deserve our attention for other reasons as well. Evidence gathered from twenty-five years of meeting minutes indicates that the group had five main goals. In addition to promoting street-tree planting and engaging in other beautification efforts, the membership urged the Park Board to make improvements to Wyman Park and other city parks and playgrounds, backed enforcement of the city's anti-smoke laws; and opposed "undesirable" commercial development in the neighborhood.

Like other improvement and "protection" associations of the day, the men comprising the membership of the Association also supported passage of the city's segregation ordinance in 1910, advocating the "enactment of such proper State or City legislation as will make it difficult or impossible for negroes, as dwellers, to invade those blocks or neighborhoods where there is a preponderance of white occupants."[41] When a U.S. Supreme Court decision struck down a similar ordinance in Louisville, Kentucky, in 1917, the ordinances of several other cities, Baltimore's included, were effectively negated as well. This did not stop the Association from engaging in segregationist practices. On April 10, 1922, noting that "no valid law can be enacted" to effect segregation, the president of the board, assured residents that "property owners in any section may by contractual agreement bind themselves not to sell to a negro." Meeting minutes from this date disclose that the association also favored protection "against invasion of the neighborhood by so-called 'kike' Jews as well as by negroes."[42]

What does segregation have to do with urban forestry? More than one might think. In reference to street trees and other forested areas, Henry W. Lawrence reminds us that "these places are often unequally available to different social groups, and as private amenities they are usually unequally distributed within the urban landscape."[43] Considering the political influence of groups such as the Peabody Heights Improvement Association, one cannot help but

Fig. 3.11. Recently pruned trees in the 500 block of South Glover Street, Baltimore, ca. 1930. Note the very small tree pits. Narrow sidewalks and inadequate tree pits were among the problems that caused high street-tree mortality during the early years of the twentieth century. Tree loss remains a problem.

wonder whether a neighborhood-wide strategy to plant trees, improve park grounds, and adopt other beautification measures, coupled with aggressive enforcement of exclusionary covenants—formal or informal—might have caused a disproportionate share of resources to flow into largely white and relatively wealthy districts such as Peabody Heights at the expense of neighborhoods of color. Evidence gleaned from the records of another neighborhood group, the Mount Royal Improvement Association, seem to support this contention.

In 1930, the Mount Royal Improvement Association published for general distribution a glossy promotional pamphlet, The Mount Royal District: Baltimore's Best Urban Section. Richly illustrated with black-and-white photographs depicting stately homes and apartment buildings, elegant hotels and monuments, and beautiful tree-lined streets, the document boasts that the "greatest achievement of the Mount Royal Improvement Association has been the subjecting of the property in its area to a restriction for white occupancy only."[44] Touting its history as "the home of old and cultured Baltimore families," the authors of the pamphlet devote special attention to the neighborhood's commitment to beautification, especially the planting of trees: "The Mount Royal Improvement Association is waging a systematic campaign for the extension of this work, and has appointed a lieutenant for each block to assist the residents in adopting a uniform plan of fencing and tree planting along the alleys and the beautifying of garages."[45] The document concludes with the claim that property values in the neighborhood had "steadily increased since they were restricted for white occupancy by the Neighborhood Corporation."[46]

To underscore the connection that existed (at least in the minds of the members of the Mount Royal Association) between white occupancy and a beautiful environment, the group printed the following message on a meeting announcement from ca. 1930: "When the present officers of the Mount Royal Improvement Association assumed office, assurances were given that plans would be presented for the maintenance of this district as the most beautiful and most desirable urban section of Baltimore, but that this could be done only after the property owners had made the district safe for white occupancy by the execution of a sufficient number of the association's protective agreements. This condition was imposed because of the impossibility of preserving, much less improving any unrestricted section of Baltimore."[47] Viewed in this light, it is easy to imagine how discriminatory housing practices, protective covenants, and the rewards of political access might have combined to produce an inequitable distribution of amenities such as parks and street trees.

Writing about the city's black community during the first two decades of the twentieth century, geographer Sherry Olson skillfully displays her gift for metaphor when she states: "The black community was a hydraulic system that siphoned a very small number into higher income and status. In the competition to enter these higher social circuits, the key strategies were complete economic adaptability, homeownership at any price, and schooling. These were the signs of ambition and status. In each circuit, the white community was able to control the valves regulating flow and pressures."[48] In neighborhoods such as Peabody Heights and Mount Royal, the white community saw to it that such valves were sealed tightly.

Reforesting Baltimore

After leaving Yale University in 1913, Brooke Maxwell set about organizing the city forester's department. Although he occupied the city forester post for only four years (he left to serve as "Curator in the Smithsonian Institute in charge of wood exhibits" and, later, as a wood technologist for a propeller manufacturer during the First World War), his contribution to the program cannot be overlooked.[49] Among other things, his responsibilities included regulating and inspecting all planting, pruning, and removal activities occurring on private property, along streets, and on public grounds; regulating and inspecting all work carried out by corporations, especially utilities and railroads; spraying trees to ward off various insects; directing management of the municipal nursery at Loch Raven; and advising the Water Board on its plan to "afforest" sections of the Gunpowder Valley.[50] An invaluable source of information concerning the early operation of this new branch of city government is the set of annual city forester reports published between 1913 and 1919.

In his first annual report to the city, Maxwell outlined his department's achievements. With regard to private tree planting, he noted that 851 trees had been planted by private individuals via the permitting system. The most popular species planted were Oriental planes (452), Norway maples (206), and silver maples (138). Gardeners and "others competent to do the work" also pruned 2,696 trees (mostly North Carolina poplars) and removed 279 dead specimens. "Wire-using companies," especially the Consolidated Gas Electric Light and Power Company, were issued permits in 1913 to prune 2,151 trees. To deal with tree planting in outlying sections, Maxwell engaged real-estate developers in "an educational campaign": "It is being indicated to these real estate men what the proper method of planting is and they are shown that the antiquated method of planting a tree in front of each house is wrong. Planting plans are sometimes made for these property developers and the Division has met with some success in having planting on private streets properly done."[51]

Next, Maxwell addressed the work carried out by Division of Forestry personnel on the grounds of schools and libraries, for individuals, and along city streets—namely, planting young trees, properly pruning desirable trees, and removing undesirable trees. Conceding it would be "impossible to comply" with the numerous requests for tree planting received by his office, Maxwell sought to maximize limited resources: "Trees have frequently been planted for individuals, who purchased their own stock. The regular planting of the Division, since it has to be very limited, is being done where the greatest number of people will be benefited. Accordingly, public drives and otherwise much used thoroughfares are selected as the most desirable places to establish trees first."[52]

In 1913, a grand total of 990 trees, some of them donated by the city of Washington, DC, were planted along Baltimore's streets. Forestry crews also sprayed poison on 2,095 trees throughout the city to fend off insects, such as the tussock moth, bag worm, fall web worm, soft scale, cottony maple scale, and oyster shell scale.[53] Similarly, Maxwell acknowledged, "it would be impossible to properly prune all of the trees in the city under several years," and so he and his assistants selected "localities where the more desirable species were growing" and focused their attention on these areas.[54] With respect to tree removals, Maxwell benefited from the assistance of Baltimore's finest: "The Police Department has co-operated heartily with the Forestry Division in making a canvass of the tree conditions throughout the city. All dead and dangerous trees were reported by the various patrolmen and these trees will be removed by the Division during the present winter."[55]

While the report for 1914 simply states that a "Division of City Forestry" and a "shade-tree nursery" were established and that "about 3,000 new trees have been planted in the city," the reports for 1915 and 1916 offer a much more detailed picture of the activities of Maxwell's unit.[56] In 1915, the Division of Forestry removed 425 "dead, dangerous, or otherwise undesirable" trees; pruned 5,300 "large valuable trees which were in decadent condition"; and sprayed 4,764 trees.[57] With respect to tree planting, the Division of Forestry planted a grand total of 1,753 trees.[58] During 1916, 757 trees were removed: 501 by the Division of Forestry and 256 by private

Fig. 3.12. Linden Avenue at McMechen looking south. Initially, the strategy employed by the Division of Forestry was to plant street trees along major routes in Baltimore and to take care of the trees the city already possessed.

individuals.[58] Meanwhile, a total of 7,313 trees were pruned: 3,194 by the Forestry Division, 1,501 by private individuals, and 2,618 by public service corporations. According to Maxwell: "The regular pruning operations of the Division were centered as heretofore on the important streets where fine old trees are standing. Some of these are: E. Monument Street, Linden Avenue, Lombard Street, Hollins Street, John Street, Roslyn Avenue, Elsinore Avenue and Maryland Avenue. The work also included much pruning around public reservations such as parks, squares and school houses. Besides the pruning of large trees on selected streets, an effort was made during the year to properly prune young trees planted by the Division, wherever this work was necessary."[59] Forestry crews employed two power sprayers to spray 10,457 trees. In addition to protecting all trees planted by the Forestry Division and other valuable trees on important boulevards "as time and funds would permit," Maxwell stated, "A great deal of spraying was also done for individuals who paid for the work at uniform standard prices."[60]

Also in 1916, 1,918 trees were planted: 1,612 by the Division of Forestry and 306 by private individuals. Of the 1,612 trees planted by the division, 753 were paid for by private individuals and corporations and four by the Paving Commission. Of the 855 plantings paid for by the division, 771 were for new plantings and eighty-four were replacements. The most popular trees for planting were Oriental plane, Lombardy poplar, Norway maple, and American elm. In a section entitled "General Tree Planting Scheme," Maxwell outlined the strategy adopted by his team: "During the year a plan was drawn for planting the more important streets of the city. The aim of this plan is to connect by important in-town streets the more important parks and squares of the city. More than forty (40) miles of streets are included in the plan, requiring a total of approximately ten thousand five hundred (10,500) shade trees to complete it. Until the completion of this plan no other gratis planting should be done by the Division of Forestry."[61]

The division spent additional time and energy during 1916 on the transfer of the municipal nursery from Loch Raven to a site provided by the Water Department at Montebello Filters. The elimination of the County Division of the Water Department meant that the Division of Forestry now assumed charge of all forestry work in the city watershed. In anticipation of this shift in responsibility, a new planting plan was developed for the remaining "unplanted" areas: "The total area covered by this plan was one thousand one hundred seventy-five (1,175) acres. Of this amount eight hundred fourteen (814) acres are yet to be planted. This area is largely abandoned farmland, lying along the Gunpowder River west of Bridge No. 2. At this time approximately three hundred seventeen thousand (317,000) trees have been planted on the water-shed, seventy six thousand five hundred (76,500) of them having been planted during 1916. The planting done during the fall season covered about fifty-six (56) acres. The greater part of the work consisted of white pine in pure planting. Small areas were planted with white ash and red oak in 50% mixture. This work was charged to the Water Department at cost plus 10%, and the cost per acre was $15.00." Due to the blight, Maxwell also worked to "dispose of all chestnut timber on the water-shed as fast as possible," particularly for use as poles, the sale of which was used to plant loblolly pine on "certain abandoned areas" at the Back River Disposal Plant.[62]

Like Besley, Maxwell was also wary of "politics." In January 1917, he informed the *Yale Forest School News* that he "had not had much interference from the politicians. This may be partly due to the fact that our Mayor was reelected last May, and we have suffered no change of political complexion."[63] Politics was always looming in the background, however. Many years later, after accepting a new position with the city government, Maxwell reported he was grappling with "some difficult personnel and political problems."[64]

Perhaps the most noticeable change registered in the 1917 report was Brooke Maxwell's replacement by Carl Schober.[65] For the most part, it appears that Schober followed the same strategy employed by his predecessor. Regular budget planting resulted in the placement of 961 trees placed on city streets. In addition, the Division planted 85,301 trees on the watershed, seventy-five for individuals, and 560 for the B&O Railroad Company. Planting for individuals occurred "in scattered parts of the City, and the expense was borne by the property owners directly benefited." The division planted another 257 replacement trees, and individuals holding permits planted 198. A total of 10,123 trees were pruned in 1917: 3,651 by the Division of Forestry and the remainder by permit.[66] The Forestry Division's vigorous attempts to combat insects continued in 1917, with some 8,871 trees sprayed.

With regard to removals, Schober's crews removed 1,116 trees, while permit-holding individuals cut down an additional 299. According to the report: "In the majority of cases the trees removed were poplars (which are considered undesirable). Unfortunately, this City possesses a large number of these trees, and they are gradually being removed at the request of property owners. . . . In many cases these trees were older established poplars along the streets which have been recently paved. In order to properly pave the streets a large portion of the root system had to be removed, thus weakening the tree and either killing it or causing it to become dangerous." Cultivation and watering, tree repairs, and extensive activity at the nursery rounded out the work of the Division of Forestry during 1917.[67] The very brief report filed for 1918 indicates that 2,051 trees were planted, 10,123 pruned, 8,871 sprayed, and 1,415 removed.[68] The following year saw yet another change in administration, with Hollis J. Howe replacing Carl Schober.[69]

The city forester's reports provide illuminating details about Baltimore's urban forest. First, they remind us that the Division of Forestry's duties were not limited to tree planting. Forestry personnel were actively involved in pruning, spraying, cultivating, and removal efforts. Other responsibilities included managing the stock at the nursery and reforesting the watershed. Second, we should remember the difficulties associated with planting trees in an urban environment. Indeed, a significant proportion of plantings every year were actually failure replacements. Third, we should recognize that the Forestry Division, confronted by numerous requests and a limited budget, had to maximize resources and prioritize requests. To this end, it focused its attention on areas "where the greatest number of people will be benefited," in the case of new plantings and in "localities where the more desirable species were growing," in the case of pruning.[70] Finally, we should take note of the work performed on behalf of private individuals and corporations who paid for the city forester's services. Did such activities favor wealthier sections of town and thus contribute to an inequitable distribution of trees? If planting on private property were driven by the ability of the resident to pay for this service (except in the case of high priority plantings along major boulevards), then one might expect to find that these activities were carried out largely in upper middle class and wealthier districts of town.

By far the biggest change to affect the Division of Forestry over the course of its short history came in 1920. The following terse statement, taken from the 1920 Board of Park Commissioner's report to the mayor and city council, says it all: "A large part of the financing of the Children's Play Ground Association and Public Athletic League has been done by the Park Department, and it has likewise taken over the Forestry Division, saving taxpayers, as the result, many thousands of dollars."[71]

Although now subsumed by the Park Department, the work of the city forester and his staff continued. Unfortunately, the annual reports that provided us with such detailed information about the Forestry Division's work were either discontinued or lost. Occasionally, references to street forestry work surface in the Park Board's reports, but they cast precious little light on the city forester's activities. Likewise, proxy sources such as the meeting minutes of the Peabody Heights Improvement Association provide us with the scantest of details. For instance, these two excerpts from the group's log for January 9, 1928, confirm that requests for tree planting were now being routed through the Park Board and that these requests were still being filled, but they tell us little more: "A communication was received From William I. Norris, President of the Park Board, stating that request for planting of trees on St. Paul St. has been referred to the City Forester. . . . Mr. Norris stated that the City Forester has planted the trees as requested in the 2900 and 3000 blocks St. Paul St."[72]

Nevertheless, enthusiasm for tree planting seemed to be on the rise as this entry from the Board of Park Commissioners report for 1920 suggests: "A new era in tree planting has been inaugurated by the Park System, and in the Arbor Day Celebration inaugurated by it in conjunction with the Board of School Commissioners, wherein thousands of school children in the city take part in the planting of trees, has proven of great educational value."[73] Similarly, a story in the May 28, 1921, issue of the *Baltimore Evening Sun* entitled "Beauties of Halethorpe Compared with Guilford's" and carrying the subtitle "Back and Front Yards Turned Into Miniature Orchards—Arborial Growths Get Expert Care," extolled the benefits of planting and caring for trees in a suburban setting. In describing Halethorpe, a suburb of Baltimore, the reporter wrote that it was:

> not the resort of multimillionaires, but a community of humble homes. . . . Encouraged by "the village forester" almost every back yard is a miniature fruit orchard. . . . Crabapple, apple, pear, cherry and peach trees give wonderful promise of well-filled pantries in winter. . . . The fruit trees take the place of vegetable gardens, as vegetables can be bought from farmers nearby more cheaply than they can be grown by the residents of Halethorpe, the majority of whom are employed in the city. In addition to the fruit trees, which make the "back yards" as charming a retreat as the front porches, the avenues are lined with maples, elms, European lindens and Oriental plains, which give distinction to Halethorpe and set it apart from any other suburb of Baltimore.[74]

The reporter credited two members of the local improvement association as well as the cooperation of the State Board of Forestry for enhancing the village's sylvan appeal:

To the indefatigable efforts of Oregon Benson, P. Dugan and the members of the Halethorpe Improvement and Protective Association credit is given by residents for the "plans" which make Halethorpe resemble Arcadia. From the time the section was first developed, about 30 years ago, especial attention has been given to landscape forestry. Indeed, it is said that Halethorpe is the only place south of New York in which such systematic care has been given to trees. In co-operation with the State Board of Forestry, Halethorpe was surveyed and every tree that is planted is placed in a spot designated in the plans drawn after the survey was completed. The trees are placed 50 feet apart, and five or six times a year, they are inspected by William H. Boyer, who is an expert landscape forester. On one avenue only silver maples are planted, on another American elms, on still another, European lindens. . . . None of the trees is expensive. Economy was aimed at in the plans, which provided that wherever a tree had already been planted in great number it would be continued along the avenue, as further development made additional plantings necessary. In addition to "laying out" the stately avenues of trees Mr. Boyer devotes considerable attention to their care. "Tree surgery" is a fine art in the hands of Mr. Boyer, who co-operates with the residents in keeping both fruit and shade trees in good condition.[75]

As the preceding passage points out, individual villages—in this case a suburb of Baltimore—could benefit from the planning and care only a trained forester could provide.

Ceding Ground

But Halethorpe was not Baltimore. With so little information available from the 1920s and 1930s, we can only speculate as to the condition of the city's trees during the years of the Great Depression and World War II. Given the times, it is unlikely that their care would have been a high priority for city officials. William Fischer, the city's highways engineer, confirmed these suspicions when he admitted that "work on trees along the streets was neglected" during this time. Writing in 1946, Fischer informed a reporter for the *Baltimore Sun* that "preliminary surveys indicate that some will need treatment and many must be razed because in their present condition they are hazards to passers-by." He also observed that "the city has a heritage of old trees planted years ago without thought to collection of types most suitable for urban planting, and that much damage has been done to footways by roots of trees whose condition makes their removal necessary." Apparently, the worst offenders were poplars, the ones planted in prodigious numbers by home developers "years ago . . . because their rapid growth assured early shade and bulk."[76] A more ominous sign that all was not well with the city's trees is found in the annual report of the Department of Public Parks and Squares for 1946. In 1946, 317 trees were planted, 4,552 were pruned, 7,483 were sprayed, and 748 removed.[77] A little more than three decades after embracing professional forestry, Baltimore was removing more trees than it was planting and was preparing, once again, to address the "shade tree problem."

Although the Maryland Department of Forestry by no means prospered during the financial crises of the 1920s and 1930s, it fared much better than its counterpart in Baltimore. Whether it was due to Besley's longevity as state forester, the experience he had accumulated, the nature of the work, or, perhaps, a combination of these is difficult to say. In any event, the state's forestry program emerged from the dark years of the Great Depression in far better shape than Baltimore's Forestry Division. It is to this second chapter of Maryland's forest history that we now turn.

Chapter Four

The Dean of State Forestry, 1923–1942

We have passed the stage when forestry was mainly a crusade. Our progress
will be measured by the degree to which forestry gets downward into the soil.
—W. B. Greeley

We are no longer so rich that we can afford to waste our heritage.
—Fred W. Besley

In his first biennial report since reorganization went into effect on January 1,
1923, Fred Besley took stock of his department. Although much work needed
to be done to reforest thousands of acres of "waste land" across the state, he
indicated that "real constructive progress" had been made among private landowners thanks to the illustrated lectures his office had delivered, the leaflets
and bulletins his staff had distributed, and the "demonstrations and personal
service" he and his assistant foresters had provided to the general public. While
fires were still "the dread enemy of the forest," Besley boasted of the "remarkably good results" that had been achieved despite insufficient funding from the
state. With respect to expanding the state forest reserves, he conceded that little
progress had been made. Yet he was quick to point out that their "value as timber" had "greatly increased" since the state had acquired them and, further, that
thousands of campers and other users were taking advantage of the recreational
opportunities now available to them. In Besley's opinion, there was every reason
to be sanguine about the future. At the same time, he was well aware of the
enormous challenges that lay ahead.[1]

In his annual report for 1926, Besley identified the three outstanding objectives of Maryland's forestry program: "The first is to stop forest fires; the second, to bring about the universal practice of forestry on state and private lands; and the third, to reclaim idle, abandoned lands and put them to work in the growing of timber."[2] One year later, he offered his blueprint for the future. To ensure that Marylanders would be less dependent on outside sources of wood in the years to come, he stressed the importance of fire protection. Although it was a problem Besley would never solve to his satisfaction, safeguarding the state's forests from destructive fires was always paramount in his mind. Next, Besley advocated embarking on a massive reforestation program. Whether it was for lining the state's roads and highways with shade trees or planting seedlings and saplings on abandoned agricultural land, Besley pressed state officials to increase the output of the state forest nursery in College Park to meet the growing demand for trees. Finally, he recommended that the state purchase large areas of non-productive forest land for the purpose of growing timber. This he deemed to be "the most important work ahead for the next ten years."[3] He believed no less than 200,000 acres, or ten percent of the state's forested area, should be transferred from private to public hands.[4]

The plan was bold but Besley was resolved to act. He knew that the work of the Department of Forestry "must be speeded up tremendously."[5] No doubt he was driven by a deep-seated desire to see his vision for the state's forests become a reality. He pictured a future in which local supply was more closely aligned with local demand, private landowners utilized the forest resources on their lands in the most economically efficient and scientifically sound manner possible, and standing forests were recognized to have multiple functions and values, from watershed protection to recreation. His passion for conservative forest management notwithstanding, he may have also been responding to criticism from some of his contemporaries that he was not doing nearly enough to grow the system of forest reserves.

Building a system of state forests would not be easy. Long before the stock market crash of 1929, the country's economy was showing signs of weakness. By the early 1930s, Americans were mired in an economic depression the likes of which they had never seen. Convinced that the crisis required a bold response, President Franklin Delano Roosevelt (FDR) tested the limits of his executive powers. Like Teddy Roosevelt before him, FDR firmly believed that government should be used as an instrument to solve problems and promote democracy. Drawing a parallel between these two activist presidents, James Chace writes that "the extraordinary display of government action in the first months of the New Deal—the famous Hundred Days—was a clear reminder that a president who believed in executive power now headed the United States once again."[6]

A weak economy was not the only problem confronting FDR. Deforestation, soil erosion, and damage due to flooding were also growing concerns in different sections of the nation. Drawing on his experience as a conservation-minded legislator from New York, where he served as chairman of the Senate's Forest, Fish, and Game Committee, FDR aimed to address the twin problems of high unemployment and a deteriorating natural environment with one broad stroke when he created the Civilian Conservation Corps, known as the CCC. Passed by Congress on 31 March 1933, the goal of this program was to provide young men ages eighteen to twenty-five with "simple work" in the areas of forestry, soil conservation, flood control, and other similar activities. Under the plan, the Department of Labor handled recruitment, the Department of War administered the CCC camps, and the departments of Agriculture and the Interior supervised conservation projects on federal and state lands.[7] Unlike the progressives a generation earlier, FDR sought to promote conservation as a means of solving a difficult social problem. Ironically, while the Great Depression produced nothing but hardship for Maryland's residents, the CCC presented Besley and his staff with a unique opportunity to advance their conservation agenda.

In many ways, the period from 1923 to 1942 was especially demanding for Besley. Droughts foiled his efforts to check the spread of wildfires. Operating budgets were slashed. Funds for the purchase of new state forests were delayed. Relations with the University of Maryland soured. Most troubling of all, some of his strategies and practices were called into question. Besley's personal life was also troubled. In 1936, Bertha, his wife of thirty-five years, passed away after a protracted illness. She died from pneumonia, but leukemia and a crippling form of arthritis had long ago sapped her strength and reduced her to an invalid. For Besley, advancing his forestry agenda in these difficult years would require skill and patience. It would test him as he had never been tested before.

Fire Protection

From the day he took office to the day he retired, Fred Besley never stopped preaching the gospel of fire prevention. "Unless the forests can be protected from fires and other destructive agencies," he wrote in 1926, "forest production must remain at low ebb, and the practice of forestry seriously retarded."[8] It was a message he and his assistant foresters sought to deliver whenever and wherever the opportunity presented itself. Unfortunately, circumstances and events largely beyond his control—some natural, some human-induced—conspired against him. While he made progress, he was never able to turn the tide completely in his favor.

In fact, records show that forest fires increased year by year throughout the 1920s and continued to confound the best efforts of the Department of Forestry well into the 1940s. Edna Warren explains why: "For one thing, the population was getting on wheels and roaming far and wide. The problem of the careless smoker was beginning to take on significance. Brushburning fires were mounting and so were incendiary fires. And more and more the origin of fires was being listed as 'unknown.'"[9] Frustrated, Besley exhorted his wardens to solve the mystery behind these "unknown" blazes.

Their findings were illuminating. In western Maryland, for example, moonshiners, huckleberry pickers, and "just plain cussedness" that caused "good-for-nothing men and boys to set fires for nothing more than excitement" were implicated.[10] Farmers also set fires to improve their pastures. To reduce the number and extent of these fires, Besley approached the problem from a number of angles. In addition to educating the public about the causes and consequences of forest fires, he sought legislative and technical solutions. He also waged an uphill battle to fully fund the fire-fighting program.

On the legislative front, Besley lobbied vigorously for amendments to the forestry act that granted to forest wardens the "powers of constables to arrest and to prosecute, along with the authority to conscript fire fighters."[11] He also implored lawmakers to adopt revisions to the brush burning law and to pass a statewide stock law.[12] Taken together, these measures invested him with the power to deal with important local issues and to implement a fire-prevention and protection strategy that stretched from one end of the state to the other. Crucial as these steps were, one could argue that Besley's greatest contribution in this area was the fire-fighting apparatus he left behind.

Maintaining that "forest protection is being reduced more and more to a science," Besley was a strong advocate of new fire towers and an improved system of communications.[13] He also placed a great deal of faith in his forest wardens and fire crews. Writing in 1940, he succinctly described Maryland's program on the eve of his retirement: "The basis for the system of fire protection established in Maryland is a series of lookout towers . . . erected at strategic points from which fires can be spotted as soon as they start, supplemented by a force of wardens who, when notified by a tower that a fire has been observed, gather together fire fighting crews and the necessary equipment, and proceed to the spot. State forest wardens are empowered by law to call upon all citizens between the ages of 18 and 50 to fight fires."[14] Although the system was not yet complete, by 1940 there were thirty-two towers scattered throughout the state.[15]

Besley also increased the number of forest wardens, expanded their responsibilities, and established a close working relationship with local volunteer fire companies.[16] New fire towers were indispensable, but so too were trustworthy men on the ground to respond to the call. In 1930, he acknowledged that the "real job of forest fire suppression" is carried out by "13 lookout watchmen, nine forest guards, 650 forest wardens, and 1,500 registered crew men."[17]

By some measures, real progress was made. In 1925, for example, Besley could point to Garrett County's stellar record of fire protection as proof that his strategy was working: "There have been no fires on the Garrett County forests for the past two years, and in fact, no fires on the state forests for several years, except on Kindness Forest which is detached from the main body of state forests."[18] Even in "normal" to "bad" fire years, Besley could find something positive to highlight. In 1932, he concluded that new fire towers and increased numbers of forest guards allowed fire crews to arrive on the scene more quickly, thus reducing the acreage consumed by individual blazes. According to the report for 1932, "The size of the average fire in 1931 was 26 acres, and in 1932, 13 acres. The 1932 record is the best for the 26 years of organized forest protection, notwithstanding the fact that there were 1,265 forest fires for the year, which is regarded as a normal year."[19] During particularly dry years, however, there was virtually nothing the Department of Forestry could do to curtail the damage caused by wildfires. The disastrous fire year of 1930 was proof of that.

Fig. 4.1. "Snaggy Mountain Fire Tower, Garrett County," ca. early 1920s. Unlike the metal fire tower shown in Fig. 2.6, this tower was constructed of wood. Notice that the tower barely rises above the treetops.

Fig. 4.2. This Volunteer Fire Department truck in Earleigh Heights, Anne Arundel County, Maryland, was state-of-the art at the time. Volunteer units such as these proved valuable in the fight against forest fires.

Fig. 4.3. Display for Forest Protection Week, ca. April 21–27, 1925. Educating the public about forest fires was a top priority for Besley and his team.

Fig. 4.4. Family Forest Day in Phoenix, Maryland, June 1, 1962. George and Richard Price receive their first Family Forest Certificate. Joseph B. Kaylor, Director of Forests and Parks (far right), and Smoky the Bear (background) join in the celebration.

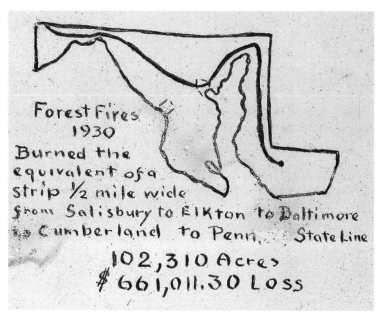

Fig. 4.5. 1930 was an especially bad fire year.

When Fred Besley remarked that "there were few days in 1930 which were not fire days," he was not exaggerating.[20] It was by far the worst fire year since the old State Board of Forestry started keeping records in 1906. The total number of fires, 2,313, was three and a half times greater than the previous high of 634, set back in 1928. The total area affected, 102,311 acres, was the largest area ever burned in Maryland in a single year. In all, there were fifteen fires over 1,000 acres in size, seven of which occurred between April 1 and May 5. The largest fire charred 7,400 acres on Dan's Mountain in Allegany County between July 28 and August 10. A 4,000-acre conflagration on Catoctin Mountain burned for five days in early May. Approximately thirty-eight percent of the fires that raged that year were triggered by smoking. Brushburning and incendiary fires accounted for seventeen percent each.[21] For Besley, the "seriousness of the drought and fire hazard" in 1930 made it "difficult to decide whether little, if any, progress was made during the year in attaining more efficient forest fire control."[22] Always the optimist, he managed to find some "encouraging aspects" despite the "great loss and expense."[23] He congratulated his forest wardens for keeping the size of the average fire, often recognized as a measure of fire fighting efficiency, below that of the most severe years during the preceding ten-year interval.

The severe drought of 1930 not only exposed weaknesses in the fire protection system, but also decimated a budget that was woefully inadequate to begin with. According to Besley, the drought was so severe that state money for lookout-watchmen, forest guards, and telephone service quickly dried up. In West Virginia (which worked cooperatively with Maryland and Pennsylvania to spot fires along state boundary lines), funding was also "quickly exhausted." One consequence of this situation was that West Virginia was forced to close its Pinnacle Rock fire tower, leaving a large portion of southeastern Garrett County with deficient lookout tower coverage. Despite the budget shortfalls, Maryland tower crews continued to report fires in West Virginia.[24] Besley also confessed that fire-fighting efficiency was diminished at times due to delays in paying fire fighters: "The counties, being unaccustomed to paying such large fire fighting accounts, quickly used up the funds provided for this purpose and delayed payments for long periods, thus discouraging the morale of the fire fighters. The Department of Forestry further embarrassed the counties through slow reimbursement of one-half of the fire fighting expense. This was because the $4,000 provided for fire fighting in the spring of 1930 was quickly obligated during the first few weeks of the season."[25]

The situation would have been much worse had it not been for the federal government. Under the provisions of the Clarke-McNary Law of 1924, the U.S. Forest Service committed to assisting states in developing and maintaining their forest protection programs. Intended as a source of funding to promote fire prevention, these federal dollars had been used in previous years for the "purchase of equipment, lookout towers, and fire fighting tools." However, with so little state money available "for meeting emergency situations such as continually arise in fire control activity," the Clarke-McNary money sometimes ended up serving as a "contingent fund" to deal with "unexpected problems."[26] In 1926, for example, an additional $600 was added "to strengthen crippled state funds" when an emergency situation arose during the spring fire season.[27] As state dependence on these federal resources steadily grew, the federal funds came to play an increasingly important part in state forestry operations. In 1926, the Department of Forestry received $4,880 of Clarke-McNary money. For the fiscal year ending June 30, 1930, that figure had risen to $10,169 with a slight increase (to $10,289) scheduled for the following year.[28] Ten years later, Besley reminded state officials just how far short of the mark their funding of forest protection work was falling: "In reply to a questionnaire by the United States Forest Service, the Department of Forestry, in 1936, estimated the annual cost of adequate protection of Maryland's forest areas at $141,501.00. . . . While this figure may be regarded as one which represents the maximum in protection called for by conditions in the State, it yet contrasts sharply with the $64,000.00 [only about half of which is carried in the regular budget] at present expended by the Department in forest protection work."[29] Suffice it to say, without this infusion of federal cash, Besley's ability to fight forest fires would have been seriously compromised.

Destructive fires continued to burn into the 1940s. In 1941, 2,045 fires scorched 46,574 acres across Maryland, with total damages topping more than $300,000. The following year, 1,527 fires blackened 39,765 acres. One year later, 1,772 fires consumed 26,209 acres.[30] By almost any measure, these were "normal" fire years. Then things started to change. In 1944, the Commission of State Forests and Parks reported just 726 fires affecting 6,484 acres and causing approximately $50,295 in damages. Compared to the average for the preceding three years, these figures represented a sixty percent decrease in the number of fires, an eighty-two percent decrease in the area burned, and an eighty percent decline in property damage sustained. How can we account for the turnaround in 1944? According to the commission, only very recently was the Forestry Department—now called the Department of State Forests and Parks—able to maintain "a properly manned and equipped fire control organization." Thanks to federal grants, Maryland was now able to dispatch "forty (40) motor-driven pumper units, with tank capacities of from 50 to 1,000 gallons," along with numerous other motorized vehicles, into the field to help douse fires.[31]

In assessing the performance of the Department of Forestry in fighting fires during the Besley years, perhaps we should pay less attention to the statistics and, instead, ask ourselves: How much damage might have been done had wildfires been permitted to smolder unchecked? This may be the true measure of the program's worth during the first four decades of the twentieth century. Not until the addition of "mobile fire-fighting units" would destructive forest fires truly come under control.[32] Until then, fighting forest fires was, and would remain, a full-time job. No number of forest wardens or fire towers could change that.

Reforestation

At the same time that Besley was attempting to defend forested areas from destructive fires, he aggressively pursued a program of reforestation. Citing a "decided reduction" in improved farmland and a concomitant increase in "waste or idle land," he suggested that much of this land could be "more profitably employed in growing timber crops."[33] Planting seedlings on tens of thousands of acres of "waste land" was a Herculean task—one that demanded a sustained effort. Moreover, it required private landowners to cooperate in large numbers in order to achieve the desired results.

Besley acknowledged this point in 1925 when he observed: "Since the forest land is practically all privately owned and divided in relatively small holdings among a very large number of individual owners, the practice of forestry rests in private hands."[34] Thus, he sought to build on the work he had begun years before, offering technical assistance to private landowners, encouraging roadside tree planting and beautification, and engaging the public whenever and wherever possible on the merits of conservative forest management.[35] Fully anticipating an increased demand for seedlings and young trees, Besley also took advantage of every opportunity to develop the State Forest Nursery in College Park.

Establishing and developing the state nursery was one of Besley's crowning achievements. Ironically, it was one of the more controversial features of the forestry program. Launched in early 1914 "for the purpose of growing and distributing at cost small seedlings or transplants for reforestation," the nursery figured prominently in Besley's reforestation plans.[36] For years, however, he complained that its output was not sufficient to meet the growing demand for trees. In his *Report for 1922 and 1923*, he bemoaned that the nursery was able to meet "less than one-third of the applications for stock," despite an increase from seven and a half to ten acres in area over the preceding two years.[37] Two years later, he commented that supply was unable to keep up with demand and that "a further extension, with increased equipment, is urgently needed."[38] Years of persistent badgering eventually paid off. Writing in 1928, Besley predicted that "in the next year . . . the nursery will reach an output of one million trees, and it will be built up in the next two years to approximately two million trees

Fig. 4.6. Getting ready to plant at the State Forest Nursery in College Park.

annually, which should be sufficient to take care of normal years even with the stimulated interest that is now shown in reforestation."[39] The following year, the nursery was, "for the first time since its establishment 15 years ago . . . able nearly to satisfy the demand for forest planting stock."[40]

Although actual production figures never approached Besley's lofty estimates during his years in office, significant gains were made. Over the two-year period covering 1924 and 1925, the College Park facility supplied more than 400 individuals or associations with 198,000 trees for forest planting and 10,435 for roadside planting.[41] In 1926 alone, 145,578 trees were distributed for forest planting and 3,722 for roadside planting, a forty-five percent upsurge over the preceding year.[42] During 1927, production more than doubled to 310,620 trees. In addition, 7,064 roadside trees were issued to interested parties.[43] Generally speaking, Besley and his staff preferred conifers or evergreens for forest planting "because they produce a more valuable product in a shorter time than do the hardwoods, with the possible exception of black locust."[44] Among the coniferous trees, white pines were favored for the western part of the state and the Piedmont, shortleaf pine for drier sections of the Piedmont and the uplands

Fig. 4.7. "Tree digger in operation, digging 9–12 ft. tulip poplars, State Forest Nursery," November 1929.

Fig. 4.8. "Tree digger hook-up at State Forest Nursery. Used in digging trees up to 3″ and 15′ in height," November 1929.

Fig. 4.9. This diorama was used to promote the value of the loblolly pine (*Pinus taeda*). Fred Besley was particularly impressed by the economic potential of the loblolly pine. After he retired, he directed the planting of tens of thousands of loblollies on Maryland's Eastern Shore.

Fig. 4.10. "Workers planting spruce seedlings by hand, Garrett County." The State Board of Forestry sought to return logged-over "wastelands" such as this piece of land to a more productive state through reforestation.

Fig. 4.11. "Co. agent W. R. McKnight with part of the cones picked at Crapo by the Crapo 4.H Club," November 1927. Young boys were paid to gather and bag the pinecones.

Fig. 4.12. "Cone Shaker–State Forest Nursery," December 1924. Using this simple rotating device, pine cone seeds were collected for planting pine seedlings elsewhere in the state.

Fig. 4.13. Silas Sines, nephew of resident warden Abraham Lincoln Sines, served
as the head of the State Forest Nursery in College Park from 1929 to 1974.

Fig. 4.14. "A group of tree wardens inspecting the shade tree nursery of the Maryland
Forestry Department at College Park, Maryland, at the second annual Tree Wardens'
Conference held at the University of Maryland."

of southern Maryland, and loblolly pine for the lowlands of southern Maryland and throughout the Eastern Shore. Red pine, Norway spruce, and European larch were also valued in certain parts of western and central Maryland.[45] The most popular trees utilized for roadside planting were elm, ash, Oriental plane, tulip poplar, red gum, and cypress.[46]

The Department of Forestry's interest in shade trees can be traced to passage of the Roadside Tree Law in 1914. This law placed responsibility for the care and protection of roadside trees throughout the state and in incorporated towns squarely in the hands of the Department of Forestry. More specifically, it was the job of a contingent of specially trained forest wardens to prevent the cutting, trimming, or removal of roadside trees except by permit. In most years, the work of these wardens consisted primarily of supervising the trimming work of pole line companies, which were chiefly concerned with securing clearance for their overhead wires. Although the law granted the Department of Forestry permission to plant trees along highways "with the consent of abutting property owners," very little of this work was actually carried out due to a lack of funding.[47]

While private citizens and town and city officials embraced the idea of planting shade trees "around the home and along the streets and highways," problems emerged that slowed the pace of progress. First, many roads proved too narrow to support such planting. Second, pole lines typically occupied the same stretches of open highway, "constituting a serious interference with proper tree development." Finally, increased motor vehicle traffic often necessitated the widening of thoroughfares, threatening the health if not eliminating altogether roadside plantings.[48] On occasion, other, more serious problems arose such as occurred in Elkton in 1930. Here, local officials contested the right of the state "to supervise the trees in incorporated towns," forcing the Department of Forestry to obtain an injunction that would prevent the cutting of trees without a permit. The case was eventually tried in Circuit Court and decided in favor of the state.[49]

Despite these problems, it is obvious that the Department of Forestry found the benefits of planting roadside trees far outweighed the headaches of maintaining them. Unlike other aspects of the forestry program, it was a commitment to aesthetics, not a preoccupation with timber production, that guided the Department of Forestry. Not only did "well-selected and well-spaced shade trees" enhance "the appearance and comfort of highways at so little expense," they could also be employed to discourage the use of "large, glaring billboards by the side of treeless highways," which formed "blots on the landscape" and marred the beauty of the countryside. According to Besley, the planting of roadside trees would "greatly reduce, if not altogether eliminate, this practice."[50]

By 1930, the number of trees distributed by the nursery for "forest and windbreak planting" had risen to 426,500. Additionally, 5,125 trees, or "enough to plant a double row of trees along 25 miles of highway," were transplanted along roadsides. Then production began to tail off.[51] Although the overall numbers for 1931 and 1932 (281,400 and 334,860, respectively) were considerably higher than those posted just five years earlier, they do show a sharp drop when compared to production figures from 1930.[52] While the Department of Forestry attributed the decline in the availability of planting stock "to the very dry season of that year . . . combined with the general depression," it is likely that other forces were at work. A passage from the *Report for 1932* offers a clue. In it Besley writes that "the distribution of planting stock from the State Forest Nursery has been further restricted to remove as far as possible any competition with the commercial nurseries of the State." He reminds readers that nursery stock intended for windbreaks "is available only for use on operating farms in the State" and only for "bona fide windbreaks," and that trees for roadside planting may only be acquired by "public institutions supported by taxation, or by landowners of the State for planting along a public highway on land upon which they reside."[53] What prompted the Department of Forestry to rein in production at the State Forest Nursery? While the general public looked favorably upon the work of the nursery, some private nurserymen were less than enthusiastic.

To put it mildly, the policies and practices put in place at the State Forest Nursery infuriated commercial nurserymen and others engaged in the sale of forest stock. They accused the Department of Forestry of trying to undercut them by offering stock at rock-bottom prices. A "recommendation" submitted by the Florists Club of Baltimore in February 1932, and accompanied by a letter from William C. Price of Towson Nurseries, complained generally of "ever increasing activity on the part of the State Forester in encroaching upon the private business of the members of this organization" and, more specifically, "against the State Forestry Department's recent advertising the sale of trees,

Fig. 4.15. Dioramas such as this one promote the State Board of Forestry's roadside tree program. Among the first states to plant and protect roadside trees, Maryland was also a national leader when it came to prohibiting unauthorized billboards on public highways.

Fig. 4.16. Roadside trees such as these provided shade and beauty in an emerging urban-romantic aesthetic, in Maryland and much of the nation.

Fig. 4.17. "Recent tree planting in Mt. Rainier. White ash from State Nursery," May 1930.

Fig. 4.18. "A typical road of Roland Park," a fashionable community near the Homewood campus of the Johns Hopkins University where Besley and his family lived. Planting and maintaining street trees was a particularly important part of the early mission of Baltimore's Division of Forestry, a legacy that is upheld today.

grown on State property at State expense and at prices which a nurseryman with taxes and other overhead expenses cannot meet."[54] In a memorandum dated 24 March 1932, Besley explained that the nursery's costs were lower than commercial producers' costs because the nursery paid no taxes, incurred no overhead charges, and specialized in only a few varieties of trees. Noting that trees were sold "at cost prices," Besley made it clear that the trees were sold for three purposes: (1) to reforest private lands; (2) to serve as windbreaks; and (3) for roadside planting.[55]

Fearing that individuals owning small parcels of land might resell trees, nurserymen wanted to restrict state sales to areas of five acres or more. The Department of Forestry countered this argument by pointing out that "there are many areas of less than five acres on farms that are subject to flood or other conditions and should be reforested." With respect to windbreaks, nurserymen expressed the opinion that farmers should buy their trees from commercial outfits in much the same way they purchased their machinery and other equipment. As a compromise, the Department of Forestry suggested that the distribution of trees for windbreaks should be restricted "to farms of 10 acres or more" and that orders should be monitored "by the county agent or a representative of the Forestry Department." Owners of commercial nurseries were also upset by advertising and publicity that emphasized the low cost of trees at the state nursery—a practice they believed fueled the perception that "prices charged by nurserymen are unreasonably high." Most importantly commercial nurserymen protested that sales to real estate companies were damaging their businesses. They demanded that shade trees sold by the State Forest Nursery only be used for planting "on state highways outside of incorporated towns."[56]

New rules governing the sale of planting stock from the State Forest Nursery were issued July 1, 1932. These rules defined key terms and offered clarification of state policies. Under the heading "Forest Planting" the rules stated that buyers had to purchase a minimum of 1,000 seedlings or transplants for planting. With respect to "Windbreak Planting," only farmers—"defined as a landowner who grows and sells from his land a farm crop, which may include live stock and poultry"—could acquire trees for this purpose. Farmers, too, were subject to a minimum purchase of 1,000 trees. Trees for "roadside planting" were available only to "public agencies supported by taxation and State-aided institutions which receive direct State appropriations for maintenance" and "landowners for planting along the highway on or adjacent to the land upon which they reside." Buyers were also required to sign an agreement affirming they would use the trees only "for the specific purpose stated." In addition, the Department of Forestry agreed to "check up annually on plantings made from trees secured from the State Nursery" and to bar violators from making any future purchases.[57]

Unfortunately, tensions persisted. In December 1934, for example, Besley received a letter from H. Street Baldwin of the County Commissioner's Office in Towson protesting against "the price list of shrubbery submitted by the State Board of Forestry as unfair competition." In his reply, Besley flatly denied that the Department of Forestry distributed "shrubbery or ornamental stock." He then dismissed any suggestion that the interests of the commercial nurseries were being compromised by the activities of the State Forest Nursery: "With 180,000 acres of waste land in the State in need of reforesting, each acre requiring about 1200 trees, and with the commercial nurseries neither able nor willing to distribute this stock at a price that waste land owners can afford to pay, I do not see how there can be any competition with commercial nurseries on this score, and I do not believe that the nurserymen themselves feel that there is."[58]

Besley then offered a vigorous defense of the roadside tree-planting program. "Throughout the suburban and residential sections of the cities and the towns," he began, "there is a keen and widespread interest in having trees planted along the highways as the most satisfactory and the most economical method of beautifying the highways and making the roadsides attractive." Considering that Maryland possessed thousands of miles of public highways, it was absolutely essential that public agencies, improvement associations, and individuals cooperate with one another "if real progress is to be made." Taking into account "that the private owner is not going to plant trees along the public highway to any extent unless extraordinary inducements are held out," trees needed to be provided at "very low cost." Besley pointed out to Baldwin that the revision of rules concerning operation of the nursery was carried out in conference with representatives of the commercial trade in 1932, and at the time "no objection was raised." He also reaffirmed his department's "genuine desire to see the nurserymen prosper" and reiterated that the state had "studiously avoided competition with their business." He concluded by arguing that the state-sponsored initiative to reforest Maryland had actually proved beneficial to commercial nurseries for "the nurserymen have been tremendously helped by our program, certainly much more than they have been damaged."[59]

Sensing, perhaps, that his reforestation efforts were being undermined or, worse, that his integrity was being questioned, Besley was more than willing to go toe-to-toe with his critics, especially if it meant advancing his forestry agenda. Sometimes his temper got the best of him. A bitter exchange of letters with William Price in 1934 only inflamed the debate over the activities of the state nursery. Frustrated that the nurserymen appeared unwilling to make any concessions, Besley intimated that the Department of Forestry would not tolerate any interference from disgruntled nurserymen when it came to carrying out their mission, warning that any "serious attempt by the nurserymen" to stop the distribution of trees "would undoubtedly react against them." Price perceived this as a thinly veiled threat and responded in kind one week later: "I do not know whether he [Besley] intends this for a challenge or not, but I am just afraid that these few words are going to make the nurserymen accept this challenge."[60] Fortunately, both parties regained their composure and nothing came of the matter.

While Besley took great pride in his reforestation efforts, nothing excited him more than to stumble upon a majestic old tree in the field. Inspired by his first encounter with the Wye Oak on Maryland's Eastern Shore in 1909, he began to photograph and document big trees across the state. Given the state's climatic and topographic variations, it was not long before he accumulated data on dozens of different species. By 1925, public interest in big trees had grown to the point where the Maryland Forestry Association (known today as the Maryland Forests Association) decided to sponsor a "big tree" contest, using a standard measurement system designed by Besley. The Association posted advertisements and awarded prizes to contestants, while the Forestry Department took measurements and compiled information. The popularity of the competition was such that the idea eventually diffused to other states. In 1940, the system of tree measurements developed by Besley was adopted nationally with minor modifications. Today, the American Forests Association maintains a National Register of Big Trees, which is published annually in *American Forests*. Besley considered the big tree program to be one of his important accomplishments, and, in 1956, an update of *Big Tree Champions of Maryland* was published.[61]

Looking back in 1932, Besley reflected on the progress of the reforestation program over the preceding decade. Conceding that there had been "a diminution of certain activities, such as the cooperative work with woodland owners" and "a slight falling off in forest planting," both due to the "condition of the times," he maintained that such backsliding was "more than offset" by increased activities in other areas. In addition to "extending and improving the forest protection system" and the increased output of the state forest nursery, Besley was referring to the tremendous expansion underway of the state's system of forest reserves, arguably his most enduring legacy.[62]

Building a State Forest System

If there was one aspect of the forestry agenda that was languishing in the 1920s, it was the land acquisition program. Although Besley proclaimed in 1923 that "state forests are the very backbone of State forestry," relatively little progress had been made in this area up to this time.[63] He acknowledged as much in 1926 when he admitted that "the acreage of state forests has in 20 years made very slight increase."[64] While he never failed to bring up the subject in his annual and biennial reports, he was unable to convince the Maryland Legislature to furnish the funds necessary to expand the system of forest reserves. That is, not until 1929, when the Department of Forestry secured its first appropriation in seventeen years for the purpose of purchasing lands. Momentum for land acquisition would actually build over the next few years, despite the bleak economic landscape of the 1930s. By 1940, Maryland boasted a state forest system that encompassed more than 106,000 acres. The path to success, however, was a circuitous one, riddled with false starts, bumps, and detours.

In his department's *Report for 1924 and 1925*, Besley repeated his assertion that Maryland should possess "200,000 acres of publicly owned forests."[65] After nearly twenty years in office, however, state forests in Maryland amounted to a mere 3,850 acres, two-thirds of which had been donated to the state. Back in 1912, the Maryland Legislature appropriated $58,500 to buy land along the

Patapsco River and to acquire "old Fort Frederick" in Washington County. According to Besley, "the purchases were made in the nature of state parks in highly developed sections where land values are high, and consequently the acreage it was possible to purchase was not large." Although the purchases were relatively small in size, Besley was pleased, noting that the lands were "meeting a high use as recreational areas."[66] In reference to the Patapsco reserve, he wrote: "The administrative policy is to preserve in a wild state of nature all the accessible portions of the state forest. There have already been constructed many miles of trails to direct travel through the woods and to make accessible the scenic points of interest. This use of the Patapsco Forest has been exceedingly popular and demonstrates in a most conclusive manner that state forests, while serving their primary purpose of timber production, may also serve other important uses, such as for camping, and the enjoyment of getting out-of-doors in natural surroundings."[67]

Since 1912, Besley's frequent entreaties to add to the state forest system had fallen on deaf ears. In 1926, Besley again made the case for expansion: "There are large areas of forest land in the mountain sections valuable for timber growing, for watershed protection, and for recreation grounds, that can be acquired at low cost—approximately $5 per acre, which under private ownership are practically waste lands." In addition to the uses mentioned above, he reasoned that adding to the state forests would "safeguard our timber supply" and "demonstrate to private owners practical methods of forest management."[68] As evidence, he pointed to the positive impact that forest wardens, fire towers, and improved communications had had on the state forests of Garrett County. Within ten years of their acquisition, fire losses had been reduced by seventy-five percent, and this section became "one of the best protected" in the state.[69] Curbing damage from forest fires, increasing productivity, reforesting waste lands, and providing recreational opportunities: these were all benefits that would accrue to the state if the forest reserve system were expanded. Convincing politicians in Annapolis to fork out tens of thousands of dollars to pay for such benefits was a more difficult sell.

Besley made yet another appeal in 1928, planning to seek approval from the legislature the following year: "The next important advance in forestry should be along the lines of extending the State Forests to include large areas of forest land, particularly in Western Maryland, that are now non-productive, but which acquired as State Forests at a low cost per acre would be built up into a source of revenue to the State, in timber production, watershed protection, and recreation areas."[70] This time, the Department of Forestry's bid for acquisition funds was successful. In 1929, Besley was able to place the adoption of a measure for purchasing new state forests at the top of his list of "noteworthy achievements." With a $50,000 appropriation in hand, he could report that "options are now held on three large tracts subject to final approval."[71]

During the seventeen-year interval that separated the 1912 and 1929 appropriations, it should be mentioned that no national forests were established in Maryland. In 1908, Maryland passed the Enabling Act that would have allowed the federal government to procure lands for national forests, but the Maryland Legislature repealed the law in 1927, much to Besley's relief.[72] Unlike his mentor, Gifford Pinchot, Besley felt strongly that public forests in the East should be administered by the states and not the federal government. He did not always hold this opinion. Earlier in his career, he supported the creation of a national forest in Maryland, going so far as to co-author a report endorsing the idea. What caused him to change his mind? Looking back, Besley maintained that, when the original law was passed in 1908, the state had no intention of acquiring "extensive areas for State Forests," and so he was willing to support the establishment of one or more national forests. With the passage of time, however, he shifted his position on the matter, increasingly convinced that the Maryland Legislature would eventually underwrite a state system. Thus, he came out strongly against the establishment of national forests, insisting that "the area suitable for public forests in Maryland is relatively small and there is not room for both." He further argued that such a stance was in keeping with Maryland's "well-known stand for State's rights and exclusive jurisdiction over her lands."[73]

To say that tension existed between those who favored the widespread establishment of national forests and those who preferred more local control would be an understatement. In 1920, the current and former chiefs of the U.S. Forest Service staked out opposing viewpoints, with W. B. Greeley asserting that "the States should be encouraged to go just as far as they will in reforestation," and Gifford Pinchot favoring a national forest policy.[74] In 1927, Greeley (who was, incidentally, a classmate of Besley's at Yale) elaborated on his position in the pages of *American Forests and Forest Life*. Contending that "we have passed the stage when forestry was mainly a crusade," he predicted that progress in the future "will be measured by the degree to which forestry gets downward into the soil."[75] Reminding readers of the leadership roles that individual states had played in instituting forestry programs prior to the period when Roosevelt and Pinchot propelled the issue to the national stage, he indicated that a "vigorous extension of state forest ownership is desirable":

> I believe that the population, financial resource, industrial interests, and public sentiment in the great majority of the states, particularly in the Eastern states, are able and ready to support a large expansion in state forest ownership, with whatever aid it may be possible to secure through county or municipal forests. And while we should go right ahead with full steam in developing fire protection, forest taxation and other encouragements of industrial and farm forestry, I doubt if there is any single item in the whole program that will give it greater strength or greater public appeal or a more specific focusing point for public action than state forest ownership on a generous scale.[76]

Greeley also believed that greater public forest ownership would "give state forest organizations the stability, the technical development, and the public standing which they need to function most effectively."[77]

Six months after Greeley's article was published, the same journal carried an editorial that was highly critical of Maryland's recent decision to repeal the Weeks Law of 1911:

> Maryland's recent action in repealing a law of twenty years' standing which gave sanction to the Federal Government, to purchase lands within the state for National Forests should it see fit, is to be regretted as an unwise and unfortunate precedent in state and federal cooperation. The action in itself has the effect of voiding the Government's authority, so far as Maryland is concerned, to carry out the long-established federal policy of forest acquisition as provided by the Weeks Act and the Clarke-McNary Act; it also is a blunt way of saying the principle of Federal cooperation in the creation of public forests does not apply in Maryland and that Maryland will have none of it.[78]

In the next passage, the editorialist takes aim at Besley himself and points specifically to the Department of Forestry's failure to build on its earlier successes:

> Just why, after twenty years, the Weeks Law has become a bugaboo to Maryland—real or fancied—to be scotched summarily, is not apparent. If the state itself were making material progress in the acquisition of State Forests the action might be more readily understandable. But such is not the case. As a matter of fact, Maryland's progress in the creation of State Forests is lamentably weak. The State was one of the first to inaugurate forestry and it has a splendid record in forest fire protection, nursery and planting practice, educational work and assistance to private woodland owners. But after twenty-five years of forestry it can show only 3,850 acres of State Forests, one half of which were donated to the State by public spirited citizens.[79]

Stating that the job of "bringing under forest management more than 100,000,000 acres of cutover lands in the South is too great" and that "the joint responsibility of both state and nation is too pressing to permit rivalry or the ghosts of rivalry" to impede progress, the writer of the column scolded Maryland for its "secession" from the Weeks Law.[80]

Besley's response, published in the following issue, sought to justify the Maryland Legislature's actions.[81] He noted that, at the time the Maryland Legislature passed the Enabling Act, "there seemed little possibility of securing money for the purchase of State Forests, and therefore a State Forest program was not contemplated." With a strong forest policy now in place, however, the state stood ready to acquire forestlands for the benefit of the public. He then enumerated the benefits that amass to states that control their own public lands: "State Forests furnish the background to support and correlate the other forest activities. They give stability to a forestry department. There is pride of possession on the part of her citizens. Their interest in forestry is stimulated because their state owns forest lands."[82] Besley also explained that federal control of land in the state would deprive its citizens of the right to tax and regulate. Were Maryland to permit the creation of national forests within its boundaries, the state forestry program would suffer:

The area of forest land requiring public ownership is not large. Should the Federal Government, at some time, in its policy of acquiring National Forests in as many states as possible, in strengthening the national programs, seek to secure lands in Maryland the State and the Federal Government would, in a limited area, be competing for the same lands and a conflict of interest would inevitably result. The State's program of extending the State Forests would be severely hampered if not entirely defeated. There would be divided responsibility in forest protection and other forms of state-wide activity. A dual sovereignty would be set up which is absolutely contrary to Maryland's stand on state rights.[83]

Letting his true "states' rights" colors show, Besley went on to sing the praises of Maryland's Governor, Albert C. Ritchie: "The State, under the able and progressive leadership of Governor Ritchie, is to an ever increasing extent realizing its obligations and serving the needs of her people. He has taken a strong stand against federal interference with the internal affairs of the State, and the fact that he has been reelected for a third term—the only man who has served more than one term as governor in the history of Maryland—shows that the people of the State are fully in accord with his policy."[84] Besley resumed his defense by placing Maryland in the company of "other eastern states which do not want National Forests within their borders," namely Massachusetts, Connecticut, New York, and New Jersey—"the very states that are leaders in the forestry movement today." He closed with a discussion of the situation in Pennsylvania, which "gave its consent reluctantly and with an understanding as to the location and extent of National Forest purchases." He pointed to the fact that, since the federal government acquired the right to purchase land, additions to the state forest system have been "almost negligible." Perhaps in an attempt to show that neither the Department of Forestry nor the state legislature took the decision to repeal the Enabling Act lightly, Besley closed his rebuttal by affirming that "Maryland is not unmindful of the signs of the times."[85]

Besley articulated his position on public lands even more clearly in a 1929 article published in the *Journal of Forestry*. A careful reading suggests he was not only wary of competition from national forests, but also mindful of the rights of private landowners. At the same time, he recognized that some level of public ownership of lands was justified.[86] Besley then posed the crucial question: "If we must have public ownership, what form is best? Should it be federal or state, or a combination?" Whether he was still smarting from the stinging criticism leveled at him two years earlier or simply defensive about the relatively small amount of land he had acquired to date, he was particularly disparaging of federal policy, especially the Clarke-McNary Act, which expanded the scope of acquisition beyond what the Weeks Law of 1911 had permitted.[87]

For Besley, the Clarke-McNary Act, while invaluable when it came to fire fighting, threatened to swamp state forestry programs with national forests. In states where forestry departments and forest acquisition programs predated passage of the Weeks Law, Besley claimed that the federal government "found the door locked against it." In states without forestry departments or with weak ones, the federal government, by contrast, was able "to enter and set up national forests," thereby diminishing the influence of the state program. Maryland, "in common with a number of southern states," had long ago passed an Enabling Act, but now "She . . . repented of her youthful indiscretion and repealed this act two years ago, before any damage had been done." National forests were a good idea, in Besley's estimation, on lands already owned by the federal government, as such was the case in the western U.S. In the East, it was "a very different matter" to "set up jurisdictions that are likely to cause friction and annoyance and interfere with the state's carrying forward a strong forest policy of its own."[88] Toward the end of his diatribe, Besley took a final parting shot at the federal program: "It is not fair that a state which in its weakness and perhaps lack of vision allowed the federal government to entrench itself as a forest owner should later, having become sufficiently wealthy and having developed a forestry department able to handle its forestry matters, be excluded from doing so because, years ago, some short-sighted official permitted the federal government to alienate its forest lands."[89]

Considering the lengths to which he carried the fight to preserve the state's authority over forest reserves within its boundaries, not to mention the prolonged drought between appropriations, Besley must have been relieved when the Maryland Legislature finally approved $50,000 for the purchase of additional lands in 1929. One senses his attitude in a letter Besley penned to Dr. Raymond

A. Pearson, President of the University of Maryland, on 17 December 1929: "I am glad to know that the Board of Regents has taken some action looking toward an expeditious handling of the land acquisition for State Forests, and you may be assured that I shall be very glad to furnish the Board with as detailed information as may be desired in setting up a policy."[90] He had been preparing for this moment for a long time.

In identifying and acquiring state forest lands, Besley favored tracts adjacent to existing state forests and "offered at a fair price" or new areas possessing the potential to grow into units of 5,000 acres or more "in a reasonable time at a fair price." Smaller parcels were also considered, if they could serve the state as "demonstration forests." Once Besley identified a suitable parcel, the typical procedure was to take out an option "for a length of time sufficient to make surveys and examine title." If everything proved satisfactory, then the option would be closed. The next step was to gain the approval of the state's attorney general, the Board of Regents, and the Board of Public Works.[91]

In a letter addressed to the Board of Regents in 1931, Besley laid out his priorities very clearly. Given the concentration of "rough mountain land, obtainable in large blocks at low cost" in Maryland's four western counties, this was where the "great bulk of state forests" should be found. Smaller demonstration forests were also desirable in southern Maryland and on the Eastern Shore.[92] In his correspondence with the board, Besley also communicated his purchasing strategy. His overarching goal was to acquire "the largest tract at the lowest possible price" in "a suitable purchase area." He placed an emphasis on obtaining low-priced lands first and recommended that the state take advantage of any opportunity to consolidate its holdings.[93]

Recognizing regional differences in land prices, Besley established purchase rates for various parts of the state where he sought to build on an existing forest or create a new one. In the vicinity of Swallow Falls and Sideling Hill, he anticipated being able to acquire land at $3.00 an acre. At $2.00 per acre, land for the Savage River and Potomac forests was even cheaper. The Doncaster and Cedarville forests were a bit more expensive at $4.00 per acre. The highest price Besley was willing to pay was $6.50 per acre to build the Pocomoke State

Fig. 4.19. Cedarville State Forest in Charles County. Although land prices were a bit higher here than in other parts of the state, Besley was interested in establishing state forests in every region.

Forest. He determined that land prices at three other locations—Fort Frederick, Patapsco, and the Seth Demonstration Forest—were too high for these holdings to be "materially increased." He also expressed a desire to acquire properties on Dan's Mountain and in the Green Ridge section of Allegany County, as well as the Catoctin Mountain portions of Frederick and Washington counties.[94]

The purpose of these new state forests was plainly conveyed in numerous documents. First and foremost, they were to be managed for timber growing. Thus, the Department of Forestry did all it could to protect these lands from fire and planted trees where needed. The state forests were also valued as protection forests "essential for water conservation in regulating stream flow, preventing erosion, and preserving the purity of the water." Finally, state forests were recognized for their recreational value, especially camping, hunting, and fishing.[95] In a brief section entitled "State Forests and Parks," part of a larger

Fig. 4.20. Old hemlocks, Swallow Falls State Forest, ca. 1958. Swallow Falls contains one of the last large stands of "old growth" forest in the state.

work called *The Forest Resources and Industries of Maryland* published in 1937 by the Maryland Development Bureau of the Baltimore Association of Commerce, Besley, elaborated on these priorities:

> The question is often asked, 'Why do States spend money for the acquisition of forest land?' There are many good reasons, but probably the first and one of the most important is to demonstrate sensible forestry practice on the ground. Experience has shown that the extension method (advising private owners), while beneficial and well worth while, does not afford the same degree of control over practical demonstrations as is generally possible when they are conducted on public lands. . . . The second reason for acquiring public forests and parks (and one as important as the first reason cited) is to make sure that the public benefits are safeguarded from private monopolistic use. . . . Another very good reason

for the establishment of State Forests and Parks is to provide for the public's recreational needs. Forests and parks provide an opportunity to hunt, fish, hike, camp, picnic, and study nature. The pursuit of these activities is becoming increasingly difficult because of the growing resentment of landowners to the public use of their lands. The only satisfactory answer thereto is the acquisition of public lands for these purposes. A fourth reason for acquiring State Forests is purely economic. The timber growth on much of our land is so depleted that private owners are not disposed to wait for another crop. As a consequence, little attention is given to such areas. Taxable wealth is thereby lost, and the more productive lands are forced to assume additional tax burdens.[96]

Stating that state forests and parks are "an excellent medium for conserving public values in the use of wild lands"—values that would likely be "lost through private exploitation"—Besley warned readers: "We are no longer so rich that we can afford to waste our heritage."[97]

During 1929–1930, the Forestry Department purchased three parcels comprising 12,312 acres. They acquired a total of 11,155 acres in two tracts adjacent to the Savage River Forest in Garrett County. The other tract, 1,157 acres in Charles County, formed the nucleus of the Doncaster Forest.[98] These were the first of many purchases to be made over the next five years. In 1931, another appropriation—this one for $25,000—was approved for "the purchase of land anywhere in the State suitable for state forests."[99] In an unpublished memorandum to President Pearson dated May 16, 1932, Besley pushed for more. Observing that Maryland, like most other eastern states, had to import lumber, he declared that it was "little short of criminal negligence, after dissipating forest resources, not to adopt constructive measure to safeguard the future." He then reminded Pearson that his "ultimate aim" was to acquire 200,000 acres.[100] Five years later, he revised these figures considerably. Noting that the original objective of 200,000 acres was "based upon surveys made some years ago," he intimated that "recent intensive studies of the forest and park needs of the State, together with an analysis of uses for available lands, indicate that the ultimate area that should be acquired is probably equal to more than twice the original estimate."[101]

By the end of the 1932 fiscal year, Maryland had acquired nearly 40,000 acres with the combined $75,000 appropriation and possessed a grand total of 48,947 acres in ten units.[102] Six additional purchases during 1933 and 1934 expanded the state's holdings even further (see *Appendix A*). Then in 1939, the Resettlement Administration handed over to the state on a ninety-nine-year lease "for incorporation into the State Forests" approximately 37,800 acres in Worcester and Garrett counties.[103]

In 1940, Maryland boasted eight state forests—Savage River, Swallow Falls, Potomac, Green Ridge, Cedarville, Doncaster, Elk Neck, and Pocomoke—totaling 106,844 acres. In addition, the state operated five parks—Patapsco, Gambrill, Fort Frederick, Washington Monument, and Elk Neck—totaling 3,619 acres.[104] Although Besley now managed a system of public lands that differentiated between forests and parks, it is clear he still favored the establishment of the former over the latter. Had he had his way, the collection of state parks under his care would have also included Rocky Gap. In 1935, Besley and District Forester Henry C. Buckingham attempted to add the "Rocky Gap tract" to the state roster. In a letter to Buckingham dated 17 December 1935, Besley wrote: "It appears that we should make a careful examination of the Rocky Gap tract belonging to the Maryland–Virginia Joint Stock Land Bank as a possible recreational area. . . . You will remember that I discussed it with you and you felt that there was not a sufficient area available in that section for setting up a State Forest, but it might have special recreational value and be developed as a kind of State Park. I shall be glad to have you look into this and let me know."[105] Rocky Gap was eventually established as a state park but not until after Besley's retirement.

On one level, Besley's bias in favor of forests needs no explanation. Besley was, after all, trained as a professional forester. Moreover, he was a utilitarian conservationist schooled in the finer points of scientific forest management. His priorities were to put a halt to destructive fires, reforest denuded land, protect watersheds, and increase timber production. But he was also interested in providing the general public with recreational opportunities. Indeed, Besley had been a consistent promoter of recreation on the state forests since the early days of the State Board of Forestry. His interest in expanding recreational

Civilian Conservation Corps workers were active in Maryland's forests and parks during the Great Depression, fighting fires and building trails, dams, cabins, bridges, and other structures. Here they are seen rebuilding Washington Monument in Washington County. **Fig. 4.21.** (left). Washington Monument, June 10, 1935. **Fig. 4.22.** (right). "Washington Monument on top of South Mountain," ca. 1936.

opportunities at the state forests is illustrated in an August 1932 letter from Buckingham to resident forest wardens in which the district forester asks for the wardens' immediate assistance regarding a new survey assessing the potential for recreational development on the state forests.[106]

The priority given to recreation distinguished a state forest from a state park: "Unlike the State Forests, the State Parks, which are relatively small, specially selected areas of natural beauty or historical interest, are intended primarily for recreation. The State Forests, on the other hand, are intended primarily for timber production and watershed protection, though, as previously shown, recreation plays no small part in their development."[107] Thanks largely to the efforts of the CCC, recreational opportunities on the state forests were expanded during the period 1935–1940. Among other things, men employed in these camps blazed trails, built cabins, and constructed two artificial lakes.[108]

Fig. 4.23. "Green Ridge C.C.C. Camp. Assembly for morning work call. Trucks with tools moving forward to take the men to work," July 1933. The CCC played an instrumental role in advancing Besley's forestry agenda. Many people today believe the CCC program should be reinstated to help curb unemployment and to improve the very infrastructure in parks and elsewhere that is a legacy of the CCC boys.

Another reason to favor forests over parks had to do with expense. Although forests far outstripped parks in terms of acreage, the total dollar amount expended for the acquisition of parks, \$66,908.87, was two-thirds that expended on forests, \$99,573.29.[109] With an appropriation in hand in 1930 and unsure whether another was ever going to materialize, Besley placed an emphasis on quantity over quality. He, therefore, concentrated his purchases in areas where he could acquire the most land for the least amount of money—even if this land had been recently cut over.[110] Viewed in the context of the acreage goals he had set for the department, it made good economic sense. Buying more expensive land along the Patapsco River or in the vicinity of Fort Frederick, near large centers of population, did not make sense at the time.

And there remained the issue of control. With pressure mounting to establish more parks, Besley grew increasingly fearful that an expanded network of parks would lead to the creation of a separate agency to oversee them. His correspondence indicates he was a proponent of parks but only if they remained under his control. This passage from a letter he wrote to President Pearson in 1935 suggests as much: "It is highly desirable that areas suitable for State parks be purchased along with State forest land. There is a strong movement for State parks, but we want to avoid a separate Park Board or Commission, which would not only add much to the expense of operation but would set up a competitive agency, probably divorced from the Forestry Department and the University of Maryland. If we obtain sufficient funds to move the park program along with the State forests, under the Forestry Department, I believe those interested would be satisfied, but unless we do have substantial appropriations for land acquisition to carry out the program, we are likely to lose out with the Park people."[111]

That same year, an ill-fated attempt to pass a bill that would have created a "State Park Commission of Maryland" drew a testy response from the state forester. In a letter to George M. Shriver, Chairman of the Board of Regents, Besley called the action "ill-advised and unjustified." Arguing that it was "an attempt on the part of the National Park Service to take the direction of park work from the State Department of Forestry" and shift it to a separate, though unpaid, commission, Besley warned that the move would result in "duplication of work" and, ultimately, the establishment of "an expensive commission, working for large appropriations in acquiring areas for parks that could be secured to the State Forest Program at much less cost." After assuring Shriver that the present administration of parks and forests was "working satisfactorily," he concluded that there was "no need of foisting on the State in these times such an expensive luxury as a State Park Commission."[112]

Three months later, Besley elaborated on his opposition to the bill in a note to Lawrence C. Merriam, Acting Assistant Director of the National Park Service. To begin with, he was disappointed that the Department of Forestry, "the present Park authority in Maryland," had not been consulted in advance. Moreover, he was deeply disturbed that the original version of the bill had been altered by the time it reached the floor. Initially, according to Besley, it was "an apparently harmless bill, carrying no appropriation." In its revised form, a $20,000 appropriation had been attached, a portion of which would have been used to pay the salary of "a director of State parks." Besley did not prevaricate. He was utterly opposed to the proposition. To avoid such a "fiasco" in the future, he advised: "If we should have additional legislation, which may be necessary, let us bring it out in the open and all parties interested cooperate to get the best results."[113]

Why was Besley dead set against the establishment of a park commission? Clearly, he viewed this as the first step toward launching a new department with a separate line of funding. Proponents of an independent park authority, no doubt fully cognizant of the state forester's uncompromising position, determined that their best hope of achieving this goal would be to circumvent the Department of Forestry. In the meantime, Besley sought to expand recreational opportunities in the state forests, hoping to forestall the movement while maintaining exclusive control over the state's modest park holdings. The result was that Maryland lost ground to other states when it came to building a separate system of state parks. Writing in 1956, Edna Warren observed that, in 1942 when Joseph Kaylor took over the Department of State Forests and Parks, Maryland was "far behind most states in providing public recreational opportunities for her citizens," or, as Kaylor put it, "It's a question now of running fast to catch up."[114]

With mandatory retirement looming, Besley could look back with satisfaction on the activities of the preceding ten years and the extensive system of public lands that had been built: a system that had once, on account of its very modest size, been a source of embarrassment. While he did not preside over the agency when it surpassed the 200,000-acre mark, he could justifiably take pride that he laid the groundwork for this eventuality. No doubt he was also pleased that forests and parks remained under his jurisdiction.

A Tight Budget

In April 1925, the *Yale Forest School News* recognized Fred Besley "as the oldest state forester in the United States in point of continuous service in one state." The alumni magazine's editors also ribbed him for being "the only state forester able to deliver a 30-minute speech in 28 minutes."[115] This gift for communication, both written and oral, proved a valuable asset when it came to winning support for his forest conservation initiatives. All too often, however, these skills were needed to defend the Department of Forestry's interests, especially its budget.[116] Never was this truer than during the decade of the 1930s. If Besley thought he had budget problems before, they paled in comparison to the austere measures he faced during the Great Depression.

Fig. 4.24. (left to right) Bill Parr, Joe Kaylor, Henry C. Buckingham, and A. R. "Pete" Bond. In the years following Besley's retirement in 1942, Kaylor, Buckingham, and Bond each took a turn running the state's forestry program.

According to one of Besley's autobiographical accounts, a key provision of the reorganization of 1923 dictated that the Board of Regents function as an independent body when dealing with matters concerning the Department of Forestry.[117] This arrangement continued until 1932, when fiscal control shifted entirely to the University of Maryland. The change had an immediate and devastating impact on the Department of Forestry. In Besley's words, "budget requests were cut mercilessly." His budget dropped from "$103,000 for general expenses and $36,000 for purchase of land and fire tower construction" in 1932 to $71,714 in 1933 and $40,290 in 1934.[118] In a letter to Major E. Brooke Lee, Chairman of the Budget Committee of the Board of Regents, Besley complained of the cuts. Pointing out that "an excellent forest protection system has been built up which has made marked progress in forest fire control," he was especially concerned that deep cuts would hinder his ability to effectively fight fires. After explaining the consequences of "insufficient" funding, he concluded: "To lay down on the job after these years of accomplishment and let the fires burn up the forests is unthinkable."[119]

Matters went from bad to worse in 1936, when a new university administration under President Harry C. "Curly" Byrd imposed even tighter controls over the budget process. In years past, Besley presented his budget to both the state budget director and the Board of Regents. Under the new administration, he was required to submit his request directly to the president of the University of Maryland, "who made such cuts as he deemed proper." Much to his annoyance and dismay, the revised version of the budget was "falsely labeled the budget request of the Forestry Department." The budget request that originated in Besley's office never reached the desk of the state budget director, nor did the Board of Regents get an opportunity to examine it.[120]

Despite the murky financial waters he was now forced to navigate, Besley continued to press for additional funds for the purchase of state forest and park lands. He justified these new requests by cleverly linking them to the activities of a popular federal program, the CCC. He argued that the state was not taking full advantage of this free labor force and that Maryland was in danger of losing it altogether because the state simply did not possess the public lands necessary to keep "the Maryland boys" busy.[121] He pursued this line of reasoning in a letter to Dr. Pearson in 1935: "I want to emphasize one point as strongly as I can because of its great importance, and that is the urgent need of money for the purchase of land for State Forests and Parks, not only because of the splendid beginning already made and the great value of these lands to the State, but more particularly to carry on the splendid program developed by the Civilian Conservation Corps camps, which are operating on publicly owned land owned by the City of Frederick. . . ."[122]

In the same letter, he told Pearson that the camps, which were operated entirely by federal funds, were a boon to the local economy and that the work that was being carried out by the camps—"the building of roads, trails, fire lines, telephone lines, lookout towers, permanent buildings, and a multitude of other things"—was adding "thousands of dollars in value of improvements to each of

Fig. 4.25. CCC workers stripping bark from logs, ca. 1930s, for use in construction projects.

the State Forests." To "secure the benefits" that were being passed on to the state, Besley maintained that "we should greatly increase the area of our State-owned lands, not only to provide work for the continuation of the camps that we have, some of which are completing their work, but also to take care of the future. Failure to do this would be a serious reflection on the State, confronted with this opportunity, and I certainly want to see the Forestry Department on record as doing everything it can to further this splendid work, which is meeting with such universal approval." At the same time Besley was asking for more money to purchase more land, he called for increased funding to ensure the "proper protection and administration" of recently purchased lands.[123]

Nine months later, Besley petitioned Governor Ritchie for additional funds to expand the state forests: "My dear Governor Ritchie, May I urge for your consideration, in connection with the special session of the Legislature which I understand is to be called, the importance in providing funds for the purchase of additional land to be incorporated in the State Forests." After assuring Ritchie of

Fig. 4.26. "Last pouring for outlet pipe, Herrington Creek Dam, Camp-S-59," October 6, 1936.

the availability of cheap lands in the southern and western sections of the state, as well as on the Eastern Shore, much of it "tax delinquent and a liability to the State," he discussed the valuable "improvement work" that has been conducted by the CCC under the Emergency Conservation Work program. Then he made his pitch to the governor:

> In order to maintain these camps in operation on the State Forests for the second six months period, it will be necessary to acquire additional forest lands. Work that can be done on privately owned land is very limited and of doubtful value so that it is exceedingly important to continue the work on State owned land, including the State game refuges. . . . Other sections are clamoring for these camps, not only for the permanent, public improvements that can be accomplished on forest land, but for the large amount of business that they bring to the respective communities in which the camps are operating. We

Fig. 4.27. "Bridge on Fifteen-Mile Creek N. of National Highway, built by Supt. H. F. Meyer with S-53 C.C.C.," September 1933. The National Road (aka National Highway) is today's US 40. It was the first federally funded interstate highway in the United States. It began in 1808 in Cumberland, Maryland, reached the Ohio River at Wheeling, West Virginia, in 1818, and was completed in 1850 in Vandalia, then the Illinois state capital. Extensions soon reached the Mississippi River at East St. Louis and Alton, Illinois.

should use every effort to retain the ten forestry camps in Maryland during the entire period of the Emergency Conservation Work activity and this does not seem possible without the acquisition of additional State forest lands.[124]

He closed with the recommendation that "$100,000 be made available for this purpose."[125] While depressed economic conditions might seriously constrain the state's budget, it also presented clear opportunities—opportunities Besley and the Department of Forestry clearly wanted to capitalize on.

In a letter to George Shriver, chairman of the Board of Regents, in 1934, Besley again invoked the CCC camps—this time, to justify the purchase of a tract on the Eastern Shore. "The acquisition of the Coulbourn tract will enable us to use the 200 men now at the Civilian Conservation Corps Camp on the Pocomoke State Forest nearby to bring it under intensive forest management without delay and this additional acreage will enable us to profitably employ the men from the C.C.C. Camp for the additional period that now seems likely, that is, until April 1, 1935." He also informed Shriver that "a large quantity of wood can be cut from this tract" and that the "thinnings and improvement cuttings" would likely "bring a substantial revenue." Maintaining that these alterations "in fact will benefit the forest," he stated that the value of the products derived from this work "will amount to more than one-third of the purchase price."[126] Two critical points emerge from this letter. First, it reminds us that Besley was very much a conservationist in the Gifford Pinchot mold. State forests, while meant to serve multiple purposes, were expected to generate income. Second, it illustrates the lengths to which Besley was willing to go to expand the forest system in Maryland. If an immediate timber harvest was needed to seal the deal, he was prepared to follow that course of action.

On the eve of his retirement, Besley continued to express his dissatisfaction with the budget.[127] At the time, he was still fuming over an incident that had occurred in 1939, when his department had been "forced to contribute from Departmental funds" some $3420.32 to help cover "a deficit incurred by the University as the result of an overestimate of probable income." Despite "urgent needs," he was irritated that the funds he requested for "minimum fire protection" were denied. He protested against the practice that prevented the "itemized recommendations of the State Forester" from reaching "the Budget Director and the Board of Public Works intact."[128] He also disapproved of the way the state relied on federal funds to meet its needs.[129] He was not in a mood to mince words: "To an extent more than appears proper," he wrote, "the Clarke-McNary and the Forest Reserve funds constitute for the Department of Forestry twin bulwarks against disaster, in that they are often employed to plug holes in an unseaworthy budget."[130]

Friction between the Department of Forestry and the University of Maryland reached a climax in 1939, when President Byrd proposed to move the Forestry Department to College Park from Baltimore, where, according to Besley, it could "be submerged under more direct control of the University."[131]

As part of the plan, Forestry Department staff members were to be made professors and assistant professors, subjecting them to university appointment. In the end, opposition from state conservation agencies as well as others "saved" the Forestry Department from consolidation with the University of Maryland. Ironically, reaction to the proposal was so strong that it sparked an effort to "divorce" the Department of Forestry from the University of Maryland altogether. A new plan called for the old Department of Forestry to be recast as one of several departments under the umbrella of a Board of Natural Resources. These departments were: Forests and Parks, Tidewater Fisheries, Upland Game and Fish, Geology, Mines and Water Resources, and Research and Education.[132] While the reconfiguration did not come with a promise of improved funding, it did have its advantages. For Besley, it meant he was no longer forced to operate under the thumb of an authority he had come to detest.

"Dean of the State Foresters"

Soon after this latest reorganization, Fred Besley, now seventy years old, retired. In 1906, he had been tapped to introduce professional forestry, a relatively new field, to the state of Maryland. Two reorganizations and thirty-six years later, he passed the mantle on to the next generation.

Did he meet the objectives he set for himself back in 1926? Not entirely. Forest fires remained a serious problem, stiff opposition from private nurserymen hampered his reforestation efforts, and he fell short of his land acquisition target. Nevertheless, in 1942, Fred Besley could look back with satisfaction on all that had been achieved. Even during the last few years of his career, Besley was at the forefront of public-private partnerships in the recreational use of state parks and state forests, as exemplified by his development of a ski resort in western Maryland. Ski areas of any kind were rare in America at the time, yet Besley saw the public good in providing for wintertime activities.

In 1931, *American Forests* trained its spotlight on Besley for its monthly "Those Among Us You Should Know" column. According to the editors of the journal, "The daily story of the forests is closely linked with names—personalities, who are pointing the way in various phases of the outdoors. Forestry, wild life and

Fig. 4.28. "Skiers relax before returning to the slope," December 1941. In response to requests from several ski clubs, Fred Besley enlisted forester Joe Davis to develop a ski resort at New Germany State Park and Savage River State Forest, both in Garrett County. Three different courses were developed between 1939 and 1941: a slope for beginners, an expert alpine run, and a cross-country course. Two private landowners—Sam Otto and Floyd Broadwater—also opened their farms to skiers. America's earliest ski resorts were just getting started at the time: Sun Valley, Idaho, in the West (1936) and Cranmore Mountain in New Hampshire (late 1930s) and Stowe, Vermont, in the East. The first ski "event" south of the Mason-Dixon Line purportedly took place here in February 1941—an affair covered by newspaper reporters from as far away as Baltimore, including the renowned photographer, A. Aubrey Bodine.

Fig. 4.29. Downhill champion Bud Little, February 23, 1941.

Fig. 4.30. "Foot of Whiskey Hollow Ski Trail—Garrett County Savage River State Forest," February 1941.

Fig. 4.31. Margaret Mast checking out the slope, February 23, 1941.

Fig. 4.32. "Watching Cross Country Ski Race New Germany Savage River State Forest Garrett Co.," February 1941.

Fig. 4.33. "Unloading for a good ski run at New Germany Savage River State Forest Garrett Co.," February 1941.

Fig. 4.34. "Coming in from the ski slope," December 1941,

Fig. 4.35. "Miss Dorothy Moore Over night sleeping Bag New Germany," February 22, 1941.

Fig. 4.36. "A little outdoor cooking New Germany Savage River State Forest," February 1941.

Fig. 4.37. "Square dance in Recreation Hall New Germany Savage River State Forest Garrett Co.," February 1941.

related fields all have their great and near great, and it is to better acquaint the public with these interesting people—men and women whose names are familiar—that this feature is conducted every month."[133] In its tribute to Besley, the article described the conditions under which he first took the job:

> Twenty-five years ago less than a dozen states throughout America had set up forestry organizations for the administration and protection of forest land within their boundaries. Among the first of these was Maryland, which organized in 1906. A Board of Forestry was created and F. W. Besley, whose tree-planting activities in the Pike's Peak area, in Colorado, had attracted considerable attention, was offered the office of State Forester. He accepted and today has the record of longest continuous service of any State Forester in the United States, having served Maryland in that capacity for twenty-five years. Under his direction, and starting from a small beginning, the Maryland Forestry Department has steadily advanced, keeping abreast of the most progressive states in meeting their particular forest needs.[134]

After recounting some of the major milestones of his career and listing his various memberships, the column concluded that Besley had now reached "the enviable position as dean of the State Foresters of the United States," a rank he would retain for another decade.[135]

Fig. 4.38. Fred Besley with his grandson, Kirk Rodgers, in the Whiskey Hollow section of New Germany State Park, ca. 1939, three years before Besley's retirement. Younk Kirk is seen here holding a dead timber rattlesnake (*Crotalus horridus*), a venomous pitviper species found in the eastern United States.

Fig. 5.1. "View from Town Hill Tower looking eastward. Sideling Hill in distance," May 1932. Fire towers offered panoramic views of the surrounding countryside. Views such as this still exist in Maryland today.

Chapter Five

Rebirth and Renewal, 1942–1968

And I congratulate our distinguished pioneer, Fred Besley on his 84th birthday, and on the trails he and his associates blazed, not only for Maryland, but for other States to follow.

—Theodore R. McKeldin

I was under the impression that all people like to have trees planted in front of their houses until I started planting trees in front of houses.

—Charles A. Young

As the United States emerged from World War II, Maryland's Department of Forests and Parks and Baltimore's Division of Forestry appeared to be headed in opposite directions. Although the State Department of Forestry had encountered its share of problems since 1906—from the anemic budgets of the early State Board of Forestry years to the turf battles of the mid- to late-1930s—overall the future looked relatively bright. By almost any measure, Maryland's early experiment with forestry was a success. After Besley's retirement, the duties of state forester passed on to a newcomer to the Maryland scene, who was, in turn, succeeded by an old hand from the Besley years. By the close of the 1960s, Maryland's forests and parks were thriving.

Baltimore, on the other hand, was struggling to maintain, let alone expand, its tree canopy in the immediate post-war years. Problems that had always plagued the Division of Forestry were now magnified. Despite the best efforts of the city forester and his staff, by the 1940s, if not earlier, it was clear that Baltimore's street trees, in particular, were suffering from disease and neglect. Faced with the difficult task of growing trees in an urban environment, the division—poorly funded and understaffed—found itself locked in a near-constant campaign just to hold the line. Indeed, by the 1950s, some officials were referring to the "losing battle" that was being waged to protect the city's trees.[1] A decade later, mounting development pressure and a growing resistance to tree planting among some citizens were complicating the already difficult job of the city forester. Only a concerted effort on the part of the Division of Forestry and the Women's Civic League, working in cooperation with city officials and other allies, as well as a critical infusion of cash, stemmed the tide of tree loss. Indeed, by the end of the 1960s, the future looked bright for the city's trees.

End of an Era

Fred Wilson Besley retired on his birthday in 1942. He was seventy years old. Five days later, he was the guest of honor at a testimonial dinner held in conjunction with the "Annual Dinner Dance" of the Maryland State Game and Fish Protective Association at Baltimore's Southern Hotel.[2] The speakers that evening included Governor Herbert R. O'Conor, Baltimore Mayor Howard W. Jackson, Acting Chief Forester of the U.S. Forest Service Earle H. Clapp, and Professor H. H. Chapman, Dean of the Yale School of Forestry.[3] That night, Besley was remembered for mapping Maryland's forest resources, working closely with farmers to manage and protect woodlands, developing an effective "fire organization," starting the State Forest Nursery, protecting roadside trees from destruction, and building a system of state forests and parks that now exceeded 118,000 acres.[4]

Those in attendance were also reminded of the many elected offices he had held during the course of his career: Director of the American Forestry Association, Secretary of the Maryland Forestry Association, President of the Association of State Foresters, member of the Executive Council and Treasurer of the Society of American Foresters, and President of both the Yale Forest School and University of Maryland alumni associations. Over the years, numerous other professional and community activities placed demands on his time, most notably the Presbyterian Church and the Boy Scouts. Besley had also received many awards, including the Honorary Degree of Doctor of Science conferred upon him by the Maryland Agricultural College in 1914 for his outstanding work in the field of forestry.[5]

Surprisingly, Besley's retirement was short-lived. Rather than "slow down" to spend more time with his four children and six grandchildren, as he had originally intended, circumstances compelled him to change his mind. His entry for the July 1944 issue of the *Yale Forest School News* explained what happened next: "[T]he war has upset my plans as in the case of most people. For the past 15 months I've been substituting for my son, Lowell, who is now a Lieutenant in the Navy. His permanent job I am now filling is associate professor of forest management, West Virginia University." To this he added that he enjoyed "working with the W.Va. group" and that he was "still going strong at 72."[6]

With more than forty years of practical forestry experience under his belt, Besley brought a perspective to the classroom that no one else on the faculty could match. In a brief article published in 1943, he reflected on the changes that had shaped the discipline since the beginning of the twentieth century. Soon after the first forestry schools were established, he recounted, the lumber business reached "astounding proportions" in what was "truly the age of forest exploitation." Having "developed the machinery for forest destruction to a higher degree than anywhere else in the world," the lumber industry moved about the country, from one forest region to the next, "leaving devastated forests in its wake." In 1909 alone, the amount of lumber cut reached forty-four and a half billion feet. Then came a decades-long decline that saw production dip to an all-time low of ten billion feet in 1932. "Silviculture and management were taught in theory," he recalled, "but with little prospect of their application in the woods." The thought that "lumber men might . . . see the light and employ these young enthusiastic foresters to their everlasting benefit" was but a "fond hope." The result was that most foresters, Besley included, entered into public service. As there were very few states that had journeyed down the road of professional forest management, this meant that most young foresters ended up working for the federal government. Maintaining that forestry's "high standards and elevated moral tone" could be traced "to great teachers like Fernow, Roth, Pinchot, Graves, and Toumey," Besley averred that "no other branch" of government had "exceeded it in public usefulness and high esteem for the past forty years." Years of practice also convinced him that, for all that forests and forestry had changed over the years, foresters "as a class" remained "as progressive and adaptable as any group of professional men."[7]

While much progress had been made, great challenges lay ahead. Besley noted, for instance, that World War II had placed significant burdens on the nation's forests, remarking that a long time would be required "to recover the damage now being done which, added to the already depleted condition of forest lands in general, creates a tremendous problem for the future." There was general agreement that public agencies needed to expand and intensify efforts to educate the general public and cooperate with private landowners to promote good forestry practice, but, in Besley's opinion, this strategy alone was "far from sufficient to do the job in any reasonable length of time."[8] What was the answer? One possibility was "public regulation of cutting," an option "urged especially by the Federal foresters as the only effective way of dealing with the present situation." More immediate action could be taken if President Roosevelt exercised his authority to issue an executive order. Besley found both of these options, especially the second one, unappealing: "Rule by decree in this democratic country of ours is odious and certainly does not seem justified in this case, in which there is no great emergency but only an accentuated condition that has been going on for many years and which has been, and is now, being considered by the Congress." Instead, Besley advocated a tried and true approach, one that funneled federal dollars to state coffers but protected the sovereignty of the individual states and the rights of private landowners to the greatest degree possible. In his view, private ownership and initiative could be maintained and forest management improved if public funds were used "as a recompense to the private owner" for managing his lands for the benefit of the general public, "whether it be for preserving a watershed cover, protecting wildlife, scenic values, or other recognized public uses."[9] The alternative was plain and simple: public ownership. However, the scale of the problem precluded the wide-scale application of this option.

Although West Virginia lagged far behind Maryland in its management of forest resources, Besley clearly enjoyed his time in Morgantown: "This group of West Virginia University foresters is a fine bunch to work with and they have made it very pleasant for me. It has taken me back to some of my pioneering days but it is gratifying to see the rapid and substantial progress that forestry is making in West Virginia." Due to the shortage of students, Besley dedicated most of his time to management of the West Virginia Forest Products Association, "a cooperative marketing and timberland management agency working with private forest land owners."[10] It was undoubtedly an aspect of forestry with which he was intimately familiar.

After three years, Besley resigned his position at West Virginia University, convinced that his colleagues there would "soon catch up to their neighboring states in most ways." He also knew that this, his "second retirement," would be "a permanent one." It was time for Besley to turn his attention to a new undertaking: management of several thousand acres of newly acquired forest lands on Maryland's Eastern Shore. "After all these years of advising others how to manage their lands," he admitted in 1946, "I am on the other side of the argument to demonstrate the practicability of such advice."[11]

Meanwhile, big changes were taking place back at the Department of Forests and Parks. Rather than replace Besley with long-time assistant, Karl Pfeiffer, Maryland tapped Joseph Kaylor to fill the position of state forester.[12] Although Pfeiffer appeared to be the heir apparent—and Besley's favored successor—there was a feeling among some that the time had come for new blood, especially in the aftermath of the dispute with Curley Byrd.[13] A graduate of the Pennsylvania State Forest Academy, Kaylor was a relative newcomer to Maryland forestry. Personable and politically savvy, it did not take him long to learn the ropes. His first task was to ensure a smooth transition for Forests and Parks as it moved from the University of Maryland to the new Maryland Board of Natural Resources.[14]

In 1947, Kaylor resigned his position to take a job with the federal government. In his absence, the office of state forester passed to Henry C. Buckingham, an experienced assistant forester who had served ably during the Besley years. One year later, Kaylor was back, this time as Director of the Maryland Department of Forests and Parks. Under this new arrangement, Buckingham reported to Kaylor but continued to serve as state forester. Offutt Johnson, whose father worked as a district forester during the 1940s and 50s and as an assistant state forester and superintendent during the 1960s and 70s, recalls that Kaylor had an arrangement with Buckingham: he worked on parks, and Buckingham "oversaw the forestry end of things."[15] Although Buckingham was aware of the "growing trend toward outdoor recreation" and understood the need to expand recreational opportunities around the state, "Buck's allegiance," remembers former Maryland State Forester Jim Mallow, "was always to the forestry side of the organization."[16]

When it came to managing the state's forests, Kaylor and Buckingham made relatively few changes.[17] "The course that Besley laid out was so visionary," notes Mallow, "that for some time no one dared to attempt to alter the course simply because it didn't need any altering."[18] Johnson, who served as Assistant Director of Program Open Space from 1966 to 1991 and then as a naturalist at Patapsco State Park from 1991 to 2001, agrees, stating that "the basic practice of the science and art of forestry remained consistent" through the Kaylor and Buckingham years. In fact, he maintains, the basic practices—tree planting, surveying, nursery work, and so on—continue right down to today: "What changed were the tools and machines. . . . [They] revolutionized forestry."[19] Nevertheless, three key pieces of legislation passed during the Kaylor-Buckingham era had a lasting impact on forest management. First, the Forest Conservancy Districts Act of 1943 "established a series of regional forestry boards to provide more localized advice and knowledge about forestry." It also enhanced the department's ability to enforce regulations governing fire and timber cutting.[20] According to Mallow, it was the first "comprehensive public regulation of forest practices on private lands" east of the Mississippi River.[21] Then there was the Mid-Atlantic Forest Fire Protection Compact of 1954, which made it possible for Delaware, Maryland, New Jersey, Pennsylvania, Virginia, and West Virginia to share resources to fight fires. Finally, the Forest Conservation and Management Act of 1957 provided incentives to private landowners to actively manage their forested acres.[22]

Content to let Buckingham handle forestry matters, Kaylor used his political skills to expand and develop the state's park and recreational facilities. From the outset, he made it clear that this would be his top priority. According to historian Robert F. Bailey, Maryland's new state forester "built on Besley's Depression-era accomplishments and helped establish 20 new state parks over the next two decades." Such tremendous growth was made possible thanks to generous support and funding from Annapolis. The allocation of new funds came about in response to several factors, including an increase in population, greater mobility and leisure time, post World War II prosperity, and a rise in demand for recreational opportunities. As a result, state forest and park attendance rose dramatically, from 240,000 in 1945 to more than 5.3 million in 1959.[23]

Although Besley did not approve of Kaylor's emphasis on parks—a point he often made to family members gathered around the dinner table—he kept his opinions largely to himself.[24] Except perhaps through his participation in professional associations or, quite possibly, through his personal friendships with former colleagues, Besley does not appear to have involved himself in agency affairs after he retired. In the judgment of Ross Kimmel, a cultural resources specialist with the Maryland Department of Natural Resources, "This was likely a conscious act on his part, now that he was managing his own timberlands."[25]

Without a doubt, Besley and Kaylor were cut from different cloth. Each brought particular talents and priorities to the table. They also faced very different challenges. Although both were trained as foresters, Kaylor presided over the Department of Forests and Parks during a period of rapid urbanization and development. It was also a time when things were "more political," recollects Mallow: "Kaylor came at a time when the public started demanding use of 'their' parks and forests." In response, he placed "a major emphasis on the development of state parks," many of which "grew out of the state forest recreation areas that had existed over time on the state forests."[26] It was Kaylor, then, who recognized the need to build a modern state park system. To this end, he created the position of Superintendent of State Parks in 1954, the goal of which was to begin to place parks on equal footing with forests. Reflecting his desire to protect forested watersheds, preserve water quality, and promote multiple use management, Kaylor urged development of parks along stream valleys such as Patapsco, Seneca Creek, and Gunpowder.

When Kaylor retired in 1964, Spencer P. Ellis stepped in to take his place. Under Ellis, Forests and Parks reached new heights: "Grand plans coupled with matching federal funds led to dramatic land acquisition and a park building spree. Beginning in the mid-1960s and lasting until the funds began to run dry in the mid-1970s, dozens of Maryland parks received modern amenities and facilities."[27] Notable among the department's accomplishments during his tenure was the creation in 1969 of Program Open Space which used tax money from real estate title transfers to acquire additional acreage for the state's system of public lands.

Fig. 5.2. Fred Besley surveying a recent land acquisition, ca. 1947.

Celebrating Fifty Years of Forestry in Maryland

Returning to Baltimore in 1946, Besley did not have to cast about for new projects to occupy his days. Upon retiring, he had acquired approximately 6,000 acres of land in Dorchester, Somerset, Wicomico, and Worcester counties on the Eastern Shore of Maryland for the purpose of growing timber. Now with plenty of time on his hands, he intended to put his forest management strategies to work. With his son-in-law, Samuel Procter Rodgers, he formed a timber company called Besley & Rodgers, Inc. Drawing on his many years of experience with Baltimore Gas and Electric Company, Rodgers's business acumen was a perfect complement to Besley's technical knowledge. Eventually, Besley & Rodgers would emerge as the largest non-industrial private landowner in the state.[28]

Within a year of returning to Maryland, Besley was ready to pronounce the new business venture a success: "I am getting some interesting experience in multiple use of forest land, privately owned," he wrote in 1947. "Loblolly pine lands, more than 95% in tree growth, purchased a few years ago purely for their timber growing value are now producing income from oil leases on a prospect basis sufficient to pay the taxes and the limited marsh areas are leased for the trapping of muskrats." In addition, he and Rodgers were exploring other possibilities for the land, including "grazing leases, holly production, [and] fishing and hunting privileges." With respect to the pine trees, they were "growing in volume at close to 10% per annum."[29] The *Baltimore Sun* noted that the strategy employed by Besley-Rodgers, "a timber company whose avowed purpose was to demonstrate that forests could be a good investment," was proving "as successful for a timber company" as it had for the state's forests and parks.[30]

Besley's budding enthusiasm for private management of forest lands even prompted him to rethink his position on state forests: "I am having to revise the argument I used before appropriation committees to get money for the purchase of state forests, that it was only in public ownership that these multiple use values could be fully developed."[31] Of course, Besley well knew that he was no ordinary private landowner. He possessed a wealth of experience that few of his neighbors could hope to match. Indeed, he was applying techniques that he and his staff at the Department of Forestry had promoted vigorously for nearly half a century.

When it came to timber resources, Besley made no effort to conceal his partiality for the versatile loblolly pine. His entry in a 1952 issue of the *Yale Forest School Alumni Magazine* documented this preference unambiguously: "He finds plenty to do combatting [sic] worthless hardwoods on his 6,000-acre tract of loblolly pine on the Eastern Shore of Maryland."[32] A few years later, he bragged that, while the farmers on the Eastern Shore were "suffering from the worst drought in 30 years," with crop production at less than fifty percent, "the pine on a 1,000 [acre] tract of swampy land we own is not only sustaining good growth but according to the two saw mill men now operating on the tract, the dry season has afforded the best logging conditions they have ever had." Besley held this up as proof of the "superiority of timber growing especially if you have loblolly pine."[33]

Besley also found time in these "retirement" years for other pursuits. He spent the entire summer of 1949 building a family cabin on Church Creek (a tributary of the Little Choptank River) near Woolford, Maryland. The loblolly logs he used in construction were cut from family land, hauled from the woods by oxen, and milled locally.[34] In January 1951, he was featured on the cover of the *Yale Forest School News*, along with A. F. Hawes ('03), W. B. Greeley ('04), H. H. Chapman ('04), W. J. Damtoft ('11), and G. A. Garratt ('23). All had journeyed to New Haven to celebrate the fiftieth anniversary of the School of Forestry. In 1956, the Maryland Forestry Association and the State Department of Forests and Parks published his sixty-page bulletin, *Big Tree Champions of Maryland*.

Perhaps the biggest highlight of his retirement came in 1956 when family, friends, and colleagues gathered again at the Southern Hotel in Baltimore to celebrate the fiftieth anniversary of forestry in Maryland, as well as his eighty-fourth birthday. On hand were Governor Theodore R. McKeldin, Chief Forester of the U.S. Forest Service Richard E. McArdle, Director of the Department of Forests and Parks Joseph F. Kaylor, Besley's twin sister, Florence, his three children, and two grandchildren. Also in attendance was Brooke Maxwell. After more than twenty years working as a landscape architect, Maxwell had returned to public service in 1945 as Director of Parks and Recreation for the city of Baltimore. From 1953 to 1961, he was also chairman of the State Commission of Forests and Parks.[35] Reminiscing, Besley recalled that he had been "reluctant to accept the job" of state forester in 1906, "fearing it might be a political one." When all was said and done, he served eight different governors, both Democrat and Republican. Among his many accomplishments, he "pointed with pride to the fact that while most states have kept their State Forests and State Parks separate, in Maryland, they have always been under the same management."[36]

In his keynote address, Governor McKeldin paid tribute to some of the great names in forest conservation history: "It was not until around the turn of the century in which we live that forestry-minded men came places where their voices could be heard by the multitudes—great conservationists like Theodore Roosevelt, who then was President of the United States—great advocates of reforestation like Gifford Pinchot of Pennsylvania—dedicated practical,

Figs. 5.3 (top) and 5.4 (bottom). Building the cabin on Church Creek near Woolford, Maryland, 1949.

Fig. 5.5. Fred Besley (right) at the cabin on Church Creek in 1949 with his son-in-law, Proctor Rogers (left), and grandson, Kirk Rogers (middle).

informed and foresighted men like Fred Besley of Maryland." Pausing to recall the significance of 1906 in Maryland's history, he continued: "Fred Besley, whom we honor here tonight, survives his fellows who were forestry pioneers. We honor their memory too. It is happy we are that we can express to Fred our appreciation for all that they did, for the trails they laid for the future and for a better State because of their accomplishments."[37] To place Besley's name among the leading lights of the forest conservation movement was surely one of the greatest compliments the governor could have bestowed upon the former grade school teacher from Virginia.

More than half a century later, Besley's legacy is difficult to pin down. To some, it is rooted in the programs that are still in existence: fire control, cooperative forest management, roadside tree planting, environmental education, outdoor recreation, and big tree champions. "I can't think of anything he started that has been discontinued," Johnson confesses. "In fact, a lot of states copied what Maryland did. . . . Besley was perfect for his time."[38] To others, including Kimmel, it was Besley's uncompromising professionalism and capacity for hard work that set him apart: "He was indefatigable. He set a very high standard for responsible civil service."[39] Looking back across the years, Johnson, too, believes that Besley set an example for all who followed in his footsteps: "Each state forester and his staff seemed to be right for their time. Kaylor, Buckingham. We never had a bad one—from Besley right down to today. Never a hint of impropriety or scandal."[40] To others still, Besley's legacy is clearly manifested in Maryland's current system of public lands. Francis "Champ" Zumbrun, Forest Manager at Green Ridge State Forest since 1978, is reminded of Besley every time he visits a state forest or park in Maryland: "What would Maryland have been like if Fred Besley had not been state forester? If he hadn't walked every cow path in Maryland?" asks Zumbrun. "We would have less state land; less forest cover on private property; less outdoor recreation. His greatest legacy is the 42% forest cover" we now possess.[41]

Especially in his later years, Besley took advantage of every opportunity to spend time "in the woods." For him it was not a pastime but a way of life. As he grew older, however, travel to the Eastern Shore became increasingly difficult.

While he still possessed a driver's license and enjoyed operating his jeep on the family property, his eyesight was extremely poor. One night in the late 1950s he drove to the cabin on Church Creek from his home in Laurel by following the lights of a yellow truck that he knew was headed to Cambridge, much to the dismay of family members. Concerned for his safety, his sister, Florence, agreed to accompany him on future trips. On one particular occasion she escorted him to an area that had been recently cut over so he could check the progress of new growth. By this time he was nearly blind. She took his arm and helped him kneel down so that he could run his fingers over the tiny seedlings. It was one of his final visits to Woolford.[42]

"A Losing Battle"

As the post-war years unfolded, three important changes affecting the care and management of Baltimore's trees took place. First, on February 28, 1946, Hollis Howe retired after more than twenty-five years on the job, "leaving vacant the post known as City Forester." After a brief interlude during which Edward Wisner "functioned" as City Forester, Charles A. Young, Jr., a trained forester educated at Pennsylvania State University, "was appointed Assistant Park Forester, assuming the duties of Park Forester at once."[43] Three months later, on May 20, 1946, Mayor Theodore McKeldin signed Ordinance No. 396, "which returned to the Department of Highways, the care and maintenance of all shade trees standing on public streets, divorcing this work from similar work within the limits of public parks." Elaborating further on the new arrangement, the Board of Park Commissioners announcement added: "The Highways Engineer has made available to the Department of Parks, a sum of money to be spent on street trees. The actual work is done by park employees under the direction of the Park Forester, and paid for by the Bureau of Highways."[44] A story in the *Baltimore Evening Sun* noted that the highways engineer would be "guided by the advice of the park forester on technical questions, such as types of trees to be planted along the streets and as to disposition of diseased or damaged trees" and that the "park forester is to be consulted on removals and treatment, as well as new plantings."[45] And finally, Brooke Maxwell, Baltimore's first city forester, accepted the position of Director of Parks and Recreation in 1945.[46] An ardent advocate of street-tree planting and an outspoken critic of unbridled development, Maxwell proved himself a key ally in the fight to protect the city's trees.

Under the enthusiastic leadership of Young and Assistant Forester Frederick S. Graves, support for the city's forestry program gained momentum. Their report for 1951 clearly and concisely offered a blueprint for the city: "In a park system the importance of trees for shade and beauty becomes more apparent each year, as many old trees are lost from natural causes, and others are lost as a result of large scale home construction projects. The Forestry Division continues its efforts to save what we have, and to rapidly add new shade and flowering trees as rapidly as funds permit."[47] Although coverage in the local papers, especially the *News-Post*, cast a generally positive light on the work of "Mr. Young's forces," the City Forester was confronted with several problems—problems that had always existed but that now combined to frustrate the plans of the division as never before.[48]

To begin with, city forestry personnel were faced with a situation whereby more trees were being removed than planted. In 1953, for example, a total of 1,202 trees were planted (268 in parks and 934 along streets) versus 1,674 removed (515 in parks and 1,159 along streets).[49] In a column for the *Sun* dated May 24, 1955, Martin Millspaugh placed the blame chiefly on a shortage of manpower coupled with a high demand: "Homeowners who like to see trees on the street in front of their houses will be glad to know that Baltimore has between 250,000 and 300,000 trees, and the city intends to keep things that way. But it is easier said than done. . . . In 1955, City Forester Charles A. Young, Jr., would like to plant 2,000 new trees, but that appears to be more than his 32-man staff can manage."[50] According to Young, requests for new trees were running twelve to eighteen months behind because of the current housing boom, budget shortfalls, and limitations in manpower. Millspaugh went on to describe the way the city handled requests for trees, informing readers that requests generally came from two sources: new housing developments and "individual home-owners in older, already built portions of the city." In the case of the former: "Mr. Young recommends . . . that residents of an entire block write up a petition, proving that everyone there wants the city to plant trees." In the case of the latter: "These are handled approximately in the order in which they are received . . . although they have a lower priority than requests from whole blocks of citizens."[51]

Another report from the *Baltimore Sun* identified additional problems. Noting that "progress and shade trees don't mix," the article blamed wider streets, narrow sidewalks, and increased carbon monoxide for the city's decline in tree numbers. As Brooke Maxwell put it: "It's a losing battle. Everything seems to hinge on speed and traffic nowadays." Maxwell then disclosed that "we're losing more trees than we're replacing." After pointing out that Maxwell "has been in the forefront of that losing battle to preserve shade trees on residential streets for many years" and that he possessed many allies in the fight, Ernest Furgurson, the *Sun* reporter, stressed that "the other side"—traffic planners, development contractors, utility firms—happened to be "armed with axes and bulldozers."[52] Maxwell was, in fact, well acquainted with the problem. In 1954, he had come out against the removal of twenty-nine linden trees along Charles Street "to speed up traffic," calling the plan "unthinkable." One year later, when the Greater Baltimore Committee proposed that twenty-five acres of Druid Hill Park be used for a new civic center, Maxwell again stepped into the fray: "Giving up acres of park property for highway extension is a bitter enough pill to swallow," Maxwell protested, "but to denude 25 acres of a world-famous park is beyond the bounds of reason."[53]

Furgurson next praised the dedication of Young but granted that the challenge was daunting: "The odds are about 9 to 1 against replacement of any tree removed in the course of a city project," acknowledged Young. "Aside from the regulations [that discourage replacement] other physical obstacles prevent tree planting in certain areas where shade is enthusiastically desired by residents. The carbon monoxide from today's traffic stunts the growth of trees in some places. The paved street beds sometimes prove to be impervious to water, and trees therefore cannot exist there."[54]

Newspaper articles and editorials revealed other problems, namely disease and neglect. Since the 1930s, Dutch elm disease and canker stain had been inflicting damage on elms and sycamores.[55] By the 1950s, their impact was being felt in many parts of the city. The *News-Post*, which had vowed to lead the fight to save the city's trees, carried stories that painted a particularly bleak picture of a city devoid of trees: "Residential streets now beautified by arching greenery would become deserts of brick and concrete. Lawns would bake under the summer sun and grass would die. Parks and squares would lose nine-tenths of their charm. . . . Baltimore without its trees would not be the same as we know it today. For this city to lose all or most of its shade trees would be unthinkable. But it could happen here! In fact, it is happening."[56]

Maxwell, still at the helm of Recreation and Parks in 1958, issued a warning at a panel discussion on roadside and community beautification at the Keep Maryland Beautiful Conference: "You folks who have been traveling about the city may have seen workmen taking down dead trees. . . . We are going to end up as a desert if we don't do something about this situation. All of us must put our shoulders to the wheel to avoid what many of us would regard as something of a municipal tragedy."[57] The next day, a *News-Post* editorial entitled "Let's Replace Our Lost Trees," urged residents to draw a line in the sand: "Today is Arbor Day, a landmark on the calendar that has special urgency to Baltimoreans this year. It is a day when we must heed the warning that the city's shade trees are dying faster than City Forester Charles Young's forces can remove them. Unless a vigorous program to conserve our trees and to plant many more is made a reality, the beautiful Baltimore of the past will deteriorate into bleak ugliness. The time to do something is now." To this end, the *News-Post* vowed "to help preserve and enhance the appearance of Baltimore as a beautiful 'green' city" and urged "all citizens and appropriate public agencies to extend full co-operation."[58]

With street tree numbers on the decline, the Women's Civic League, which had worked so diligently to see that Ordinance 154 was passed back in 1912, took action. Working with city agencies, the group sought to "prepare a long range program for the planting and maintenance of shade trees," curtail the destruction of trees by contractors, and follow through with a plan for planting 1,000 flowering crabs in Baltimore per year for ten years. According to the group's president, Mrs. Russell Wonderlic, any plan for rebuilding downtown Baltimore "should include trees. . . . To Keep Baltimore Beautiful it is necessary for us to preserve the trees we have and to plant new ones. I think an educational program could make the public more conscious of our need for shade trees and enable all of us to be of more help to the Bureau of Parks and the city forester in a "Keep Baltimore Beautiful" campaign."[59]

City officials also got into the act. In 1958, Mayor D'Alesandro appointed a five-man committee to study the street-tree problem. The group consisted of S. Lawrence Hammerman, Park Board Vice-chairman; George A. Carter, Director of Public Works; Charles Hook, Superintendent of Parks; R. Brooke Maxwell, Superintendent of Recreation; and Charles A. Young, City Forester.[60] Among other things, their report revealed that no fewer than 500 of the city's 225,000 street trees were dead and required removal and that none of the city's one million trees had been fertilized during 1955 or 1956. On June 9, the mayor's Committee on Trees recommended a five-point program to beautify the city.[61] An editorial in the *News-Post* applauded both the mayor's actions and committee's recommendations:

> Action to achieve a more beautiful Baltimore! That prospect is implicit in the prompt and emphatic recommendation of the mayor's Committee on Trees for a sweeping program of tree planting and conservation. It includes extensive block plantings, prunings, improved disposal of dead trees and brush, and mist blower sprayings to combat destructive insects. A special program for downtown tree plantings with cooperation of property owners is a gratifying feature. The report stems from efforts of this newspaper to promote restoration of Baltimore as the "city of a million trees." It reveals sad neglect of the city's arboreal beauties over a long span.[62]

Encouraged by the news, residents of Canton invited the mayor and city forester Young to visit the eastside neighborhood in July. But there was a string attached. "We want him to face the music and promise us some trees," said Mrs. Gustav Kallman, the president of the local council. "If he will agree to plant them it will be the sweetest music we've heard in a long time. We want the city forester there to hear it."[63] A recent survey conducted by the *News-Post* had turned up just 103 trees in an area covering sixty city blocks.

Between 1957 and 1965, the ratio of tree plantings to removals fluctuated tremendously (see *Appendix B*). In 1957, for example, twice as many street trees were planted as were removed, while in the parks more than three times as many were planted as removed.[64] The following year saw a complete reversal, with the total number of trees removed greatly exceeding the number planted. Both the impact of the March 19 snowstorm, which "dominated the activities of the Forestry Division during the year," and the "economy measures instituted by the administration," which curtailed street tree planting, contributed significantly to the drastic shift.[65] In 1959, total plantings again outpaced removals, although more trees were removed than planted in the city's parks. The report for 1959 also notified readers that the "census of trees on the public highway[s] that was begun in 1956 was concluded during the year." Cautioning that the "count is not final," the report announced: "On July 1, 1959 the census of trees were as follows: elms 9,364, maples 28,370, oaks, 2,039, planes, 13,960, ashes 2,045, linden 4,148, poplar 3,041, mixed species 5,647" for a total of 68,614. Significantly, the report determined that 4,650 blocks did not contain any trees.[66]

Plantings surged ahead of removals again in 1960, but the gain was only temporary.[67] In 1961, the pendulum swung back in the other direction. "As the total annual loss of trees through all agencies exceeds an estimate of 2300 trees, it is necessary," Young advised, "to plant a similar number of trees to maintain the tree population. An annual planting of 3000 trees is desirable to keep pace with the expansion of residential and park development."[68] Even taking into account additional plantings carried out in cooperation with the Baltimore Urban Renewal and Housing Authority, tree removals far outstripped plantings for the next three years.[69] Among other things, drought and vandalism were cited as factors contributing to the drop off.

Try as it might, the Division of Forestry could not keep up. In 1965, city councilman Thomas Ward submitted a plan "to plant more than 8,000 trees along Baltimore streets this year"—more than four times the number planted the preceding year.[70] While Ward's goal was not attained, figures for 1965 nevertheless show a "banner year" for tree planting: 6,679 trees planted for a net gain of 4,593 to the tree inventory. To finance the effort, the Board of Estimates appropriated $150,000 from general funds and $150,000 from "Gasoline Tax money apportioned to the Bureau of Highways." Actual tree planting was carried out based on a "Master Planting Plan for the City of Baltimore" produced by City Forester Graves in response to an ordinance passed by City Council. The plan divided the city "into 14 segments based on similar requirements and environment."[71]

Fig. 5.6. Urban renewal projects, such as the one carried out in the Mount Royal neighborhood during the late 1950s and early 1960s, included street trees and street lamps in the rehabilitation plan. This image is of 253 Robert Street.

According to Graves's calculations, "the cost of properly upgrading present street plantings in thirteen of the fourteen described areas would range from $14,775 to $88,670. In the remaining one, a section of East Baltimore, the cost would be nearly $385,000, however." Graves, who took over the city forester position in 1962, described the area as "practically denuded of trees." By contrast, the area of the city "with the lowest estimated cost of proper tree upgrading"—a section of town bounded by Echodale Avenue on the east, Frankford Avenue on the north, Hillen Road and Erdman Avenue on the west, and the city line to the south—possessed trees on ninety percent of its streets.[72] There could be no denying that East Baltimore stood apart from the rest of the city when it came to tree coverage.

Clean, Uncluttered Concrete

A difficult reality for forestry professionals to accept was the fact that some Baltimoreans objected to having trees planted in their neighborhoods. In the past, some people questioned tree planting on the grounds that it was too expensive, but by the 1950s it was becoming increasingly clear that cost was not the only consideration.[73] Rather, they disapproved because they simply disliked trees. In a 1955 interview, Young remarked: "I was under the impression that all people like to have trees planted in front of their houses until I started planting trees in front of houses."[74] One year later, *Sun* reporter Ernest Furgurson confirmed the phenomenon when he wrote: "And, hardest of all for tree-lovers to understand, there are some human beings who heartily object to planting projects in their neighborhoods. Such opponents, who base their attitude mostly on the fouling of underground pipes by tree roots, are the persons who make petitions for planting necessary. Mr. Young explained that it was most discouraging to attempt to plant a nice sapling before someone's home, only to draw loud protest."[75]

Some years later, the aforementioned survey conducted by City Forester Graves revealed that East Baltimore was almost devoid of trees, yet residents there were adamantly opposed to tree planting. A *Baltimore Sun* account, titled "Anti-Tree Rebels Prefer Concrete," summarized the attitudes of residents:

"I just don't like trees," Mrs. Joseph Lisiecki said. "I don't like greenery. I like clean, uncluttered concrete." Mrs. Lisiecki of 810 South Robinson street, was talking about the city's tree-planting program in East Baltimore. "If I wanted a tree, I'd move to the suburbs," Mrs. Lisiecki said. "I live in the city. I love the city. We have parks. It's nice to look at a tree once in a while. But I don't want one in front of my house." Mrs. Lisiecki's anti-tree stand is by no means uncommon in East Baltimore. "Yes," said Fred Graves, city forester, "We have met a lot of resistance in East Baltimore. Sometimes a whole block turns us down." "These people don't pull any punches," he said. "They tell you that if you want to plant trees, plant 'em in the country, not over here." "When we started large-scale planting I was surprised that so many people don't like trees," Mr. Graves said. "I was amazed." "They are old, established city dwellers," he said.

"No Johnny-come-latelies. They've been in the community for years and years."
"People say their families have been there three generations. They say: 'My
father didn't like trees, my grandfather didn't like trees, so I don't like trees.'
"It doesn't cost them a cent," Mr. Graves said, his amazement still apparent.
"It's absolutely free." "I think it's a good idea, planting trees there," Mr. Graves
said, a little wistfully. "East Baltimore is practically detreed." And that is just
the way a great many East Baltimoreans like it.[76]

The article proceeded to list the reasons why East Baltimoreans were resistant
to tree planting:

"They'll never put a tree in front of my house," Louis Averella, of 530 North
Linwood avenue, vowed. "I think trees are a nuisance in the city," Mr. Averella,
president of the Greater Baltimore Democratic Club, said. "Trees belong in the
country, not the city. If anyone wants a tree in front of his home, let him go and
live in the country." "As far as beautifying the city," Mr. Averella said, "Let's do
away with unnecessary dead trees that look lousy throughout the neighborhoods
of Baltimore." "Mrs. Lisiecki . . . said she didn't like trees because she would have
to rake up leaves." "And the juice gets all over the cars. Catterpillars [sic]—from
the trees—go up the wall. Nobody wants to be bothered with that," she said.
"You know when the wind blows," she said, "they kind of brush against your
windows. That would drive me nuts." Mr. Averella opposes trees because of the
"nuisance of birds on the cars." "When trees grow, they crack the pavements," he
added, "which have to be repaired at the homeowners' expense."[77]

Frustrated, Graves tried in vain to explain that the trees he and his assistants
planted were virtually trouble free, although he did have to allow that they still
possessed leaves:

Mr. Graves said these were common complaints. . . . "We try to tell them the
trees we plant don't do that," the municipal woodsman said. "We plant 'tailored
trees.'" Tailored trees, he explained, are small trees developed in the last ten

years for street planting. "They are practically trouble-proof," Mr. Graves said. "No problem with curbs, sewers or sidewalks." "They do have leaves," he admitted. But his arguments are apparently unconvincing to great numbers of East Baltimoreans. The forestry department surveyed the 3500 block East Fayette street and found that there was room for thirteen trees but only five locations were okayed by the residents. In the 3600 block East Fayette, the foresters found room for seventeen trees. They planted three. In the 200 block North Ellwood, the survey showed the treemen could plant thirteen trees. Homeowners cancelled out seven. In the first block North Potomac, it would have been possible to plant seventeen; so far twelve homeowners have indicated they prefer concrete. "This is pretty much the pattern in East Baltimore," Mr. Graves said. In other sections, anti-treeism is not so pronounced.[78]

One capricious editorialist offered this tongue-in-cheek solution to Baltimore's shade tree problem in May of 1967:

This is the city of Pine street and maple avenue; of still other thoroughfares whose names honor the mulberry, the elm, the dogwood, the chestnut, the cedar, the hickory—the good woods individually and the forest collectively. But this is a city with an anti-tree faction. Walk the rowhouse streets of East Baltimore, where the paving stretches smooth and clean from curb to house wall, and inquire of the average resident whether he or she would like for the city forester to plant a free tree in front of his or her house, and prepare for a firmly negative answer. . . . What the city forester should perhaps offer instead is a new and different strain of sidewalk tree, one of any desired height and foliage circumference, one whose entire upkeep will consist of an occasional dusting off, a tree that only the chemical synthetist can make.[79]

While some in the Forestry Division might have failed to see the humor of installing plastic trees on city streets, others—weary from the never-ending battle to plant and maintain trees in a harsh, sometimes hostile, environment—must have thought twice before discarding the idea altogether.

How can we account for the negative "anti-tree" attitudes of East Baltimoreans at this time? Some scholars posit that there may be a cultural explanation. According to Evan Fraser and Andrew Kenney, there appear to be sharp differences of opinion with respect to trees among different cultural groups. Relying chiefly on interviews with residents in four distinct cultural communities in Toronto as well as vegetation inventories and archival information, their research indicated that British residents reacted the most favorably to shade trees and were the most likely of the groups studied to plant a tree on their property. Members of the Chinese community, on the other hand, exhibited a preference for landscapes devoid of trees and were the least likely to add a tree to their property. Members of the Italian and Portuguese communities displayed a partiality for vegetable gardens and fruit trees and looked unfavorably upon shade trees if they interfered with their gardening activities. In another study, John Dwyer, Herbert Schroeder, and Paul Gobster found that urban African Americans, generally speaking, are not "heavy users" of forested areas. Their research on African-American communities in Chicago revealed the "persistence of negative images of forests in rural areas of the South, where many have their roots," with concerns ranging from the presence of insects and snakes to "threats of life and limb." Female children, in particular, exhibited "strong fears of forest environments." James Lewis and Robert Hendricks have reported similar findings.[80]

Does this explain why the residents of East Baltimore, largely of southern and eastern European descent, cared so little for trees in the 1950s and 1960s? While difficult to prove, cultural preferences and traditions may play an important role in shaping how particular groups perceive the urban forest. If nothing else, the matter merits further investigation.

The resistance of East Baltimoreans notwithstanding, street trees and professional forestry were here to stay in Baltimore. In spite of the budget shortfalls, the physical obstacles, and the opposition in some quarters, a solid foundation had been laid by the likes of Maxwell, Howe, Young, Graves, and scores of others. By the mid-1960s, the urban forest had also attracted renewed attention and funding from government officials and garnered critical support from residents and groups representing a broad spectrum of interests from across the city.

Chapter Six

A Rich Inheritance, 1969–2006

The young woodland remembers
the old, a dreamer dreaming
of an old holy book,
an old set of instructions,
and the soil under the grass
is dreaming of a young forest,
and under the pavement the soil
is dreaming of grass

—Wendell Berry

In 1912, Gifford Pinchot commented on the "constantly increasing activity in forestry" he was witnessing at the state level. He noted that "only a few years ago not over half a dozen States employed trained foresters," but now "more than 20 different States have some kind of organization for forest work."[1] He then proceeded to name those states that had officially established professional forestry programs. Nationally, Maryland was the third state to hire a trained forester, placing it at the vanguard of the state forestry movement. It was also the first of the Southern states to embrace formally the principles of scientific forest management.

Under the direction of Fred Besley, Progressive Era-style resource management became entrenched. While aesthetics and recreation were part and parcel of the overall program, it was fire protection, reforestation, and increased timber production that received top priority from the state forester and his staff. Only after 1942, when mandatory retirement forced Besley to step down, was greater emphasis placed on growing the state's system of parks. Besley's cautious approach to park building aside, in historian Jack Temple Kirby's estimation, Maryland's experience with professional forestry "must be deemed a huge success, and a model for the lackadaisical commonwealths to the south." In the face of "continued urbanization, then runaway suburbanization," he observes, "Maryland's forests in 1980 were almost exactly the same size as in 1910—somewhat more than two million acres." But development pressure and "hard economic realities" have taken a toll over the past thirty years.[2] Disease, the spread of invasive exotics, and global warming only add to the worries of present-day resource managers.

In Baltimore, it was local citizens, not a cadre of concerned conservationists, who galvanized the forestry movement. Groups such as the Municipal Art Society, the Women's Civic League, and a host of neighborhood improvement associations saw to it that the city hired a professionally trained forester to plant and maintain the city's trees. While air quality and shade were important considerations, a deep-seated desire to bring nature back to the city drove them to act. Through the years, these and other groups have fought vigilantly—albeit with mixed success—to ensure that the city's initial investment in park and street trees was not squandered. As Gary Moll laments, however, most of us "know the value of a board better than we know the value of reducing storm water flow, controlling heat island temperatures, or dealing with poor air quality."[3] Since the late 1960s, insufficient funding and erratic support from local residents have contributed to a decline in canopy cover. Perhaps this explains why Baltimoreans found themselves in the all-too-familiar position of having to embark on yet another major tree-planting campaign.

Old Problems, New Challenges

In 1965, things were looking up for the forests of the Old Line State. Maryland was rapidly expanding its system of parks, Baltimore was planting thousands more trees than it was removing, and money and support continued to flow in from the nation's capitol, the state house, and City Hall. Ten years later, budget cuts, development pressure, and shifting priorities were beginning to stall progress. By the 1990s, low morale and a shortage of personnel could be added to the growing list of troubles.

During the first half of Spencer Ellis's term as Director of Forests and Parks, anything seemed possible. State and federal funds were plentiful, and the department thrived. In 1969, the state underwent yet another reshuffling, creating a Department of Natural Resources under which all of its natural resource agencies were organized. Three years later, Forests and Parks was split into two separate entities: the Maryland Park Service and the Maryland Forest Service. Land acquisition, leasing, and facility development was shifted to a new unit, the Capital Programs Administration.

As the decade of the 1970s progressed, "hard economic realities . . . forced a reevaluation of Ellis's ambitious development plans." Money was becoming increasingly scarce. "As budgets dwindled," writes Robert Bailey, "most of Ellis's plans were scaled back."[4] Under State Forester Adna R. "Pete" Bond, who replaced Henry Buckingham in 1968, and William A. Parr, head of the Maryland Park Service, "maintaining existing facilities and making do with less" became the order of the day.[5] For the state foresters who followed in Bond's footsteps— Donald MacLauchlan, Tunis Lyon, James Roberts, John Riley, James Mallow, and Stephen Koehn—it would become an all-too-familiar refrain.

While Maryland possesses a great legacy in state-owned forests and parks—around five percent of the total area—only rarely have those tasked with managing this resource been provided with sufficient funds. "Budgets get smaller and smaller and the staff gets smaller and smaller," laments Offutt Johnson, the thirty-five-year state park veteran. "Right now it's so critical that they're just hanging on. Real resource management and education are not going on. They're just dealing with the day-to-day crises." Preventive maintenance, interpretation, and educational programs for school youth groups (if they are done at all) have been handed over to volunteers. "While volunteer efforts are sincere these efforts are minimal when compared to full-time professional programs of the past. . . . There was only one time when the two agencies got close to what they needed," Johnson recalls, "and that was in the 1960s." Reconciled to the fact they are never going to have enough money, forest and park managers supplement their budgets through donations and volunteer friends of the park groups. "There ought to be a way to fund state parks so they reach their full potential," muses Johnson. "They're really unique. A gift Marylanders pass on to future Marylanders. They'll always be nice, but they'll never get up to their full potential until some of these budgetary roadblocks are taken away."[6]

"When I think that Fred Besley had a staff of four people for the whole state I try not to complain," jokes Francis "Champ" Zumbrun, Forest Manager at Green Ridge State Forest. "Am I optimistic?" he asks rhetorically. "Yes. Maryland has been blessed with a long line of visionary conservation leaders." He nevertheless believes that the state is at "a critical stage" right now: "We're developing 14,000 acres a year. We need to establish smart growth principles to keep growth in places where there is existing infrastructure." As Maryland continues to urbanize, greater demands will be placed on the state's public lands. "People are coming out here to enjoy the green spaces. These people are going to need places like Green Ridge State Forest and Savage River to recreate."[7]

Jim Mallow, who held down the position of state forester from 1995 to 2001, is more pessimistic. He believes that the battle for Maryland's forests will be won or lost on private land: "The economics of growth and the resulting expansion of the population will ultimately swallow up the thousands of forested acreages and farm land of Maryland to the point where the management of the remaining forests and farmlands will not be of the magnitude and scope necessary for effective and efficient forest management and sustainability."[8] In addition to private forest fragmentation, the state's forests—both public and private—are threatened by a wide range of invasive species and forest pests.[9] Resource managers must also manage deer populations and monitor ATV (all-terrain vehicle) use.[10]

Kirk Rodgers, grandson of Fred Besley and the president of the family timber business, agrees that the threats are real and, further, that private landowners may hold the key to a more secure future for Maryland's forests. "Rapid urban and suburban development is overwhelming Maryland's landscapes," affirms Rodgers. "[T]his battle of development versus environment is going to be mostly fought out on private lands which constitute seventy percent of the remaining forest in the state. Government will play an important role through its policies on taxation and incentives for keeping forest as forest, but the key decisions will be in the hands of the 130,000 private forest landowners in the state." He, too, agrees that the problems are enormously complex, citing global warming as "a central concern." Loss of forest cover not only limits the capture of nutrients and pollutants which threaten Chesapeake Bay, warns Rodgers (who spent thirty-six years working in the environment and development field for the Organization of American States), but also reduces the sequestration of carbon from the atmosphere.[11] With regard to the future, the third-generation Yale School of Forestry graduate is not ready to concede defeat:

> Grandfather fought the twin threats to forests in his era, which were fires and over-harvest. Today urban and suburban sprawl destroys forests in a far more decisive and permanent manner. The actions he took to control fire and improve management practices seem simple by comparison to what is ahead for today's managers. But I am cautiously optimistic. The public is catching on to what is happening. The negative consequences of uncontrolled development and the loss of forest cover will probably force some changes in both public policy and private action. Grandfather if he were alive today would probably be motivated by these challenges.[12]

The challenges are many but so are the accomplishments. Despite the fact Maryland's population has more than doubled since 1950, the state maintains half a million acres in public lands. Keeping state forest land safe from encroaching development and minimizing the impact of development on privately owned forest land will take more than vision, however. It will take the cooperation of Marylanders from a diversity of backgrounds—farmers, urban dwellers, suburbanites, timber growers, woodlot owners, resource managers, and environmentalists—each with a place at the table, each dependent on a healthy forest, each willing to chart a course towards a sustainable future. All parties must be willing to base their land-use decisions on sound science and long-term objectives, not instant gratification and short-term profits. All parties must be willing to work together to tackle difficult problems such as urban sprawl and climate change. All parties must be willing to reform a system that encourages the development of a wooded acre in rural Maryland but discourages the redevelopment of a vacant parcel in Baltimore. Without this sort of cooperation and commitment, the long-term health of Maryland's forests cannot be assured.

Fig. 6.1. Meeting the challenges of the future in forestry has always required the cooperation of private landowners. Here, Louis Goldstein, Comptroller of the State of Maryland, dedicates a tree farm as part of the American Tree Farm System (STFS). Since its founding in 1941, ATFS has certified (as of 2010) 124,000,000 acres of privately owned forestland and more than 90,000 family forest owners in forty-six states, each owner committed to sustaining forest stewardship and healthy watersheds and habitats for wildlife and plants.

Baltimore's recent forest management history closely mirrors that of the state. Calvin Buikema, who served as City Forester from 1969 to 1982 and then as Superintendent of Parks from 1982 to 1996, has witnessed firsthand many of the changes that have taken place over the past forty years. "When I first came here in '69, things were really good," remembers Buikema. "We were planting 5,000 trees a year. We had a lot of money and a lot of men." By the mid-1990s, however, the city was contracting a lot of work out. There was a lack of money and men and the contract workers couldn't fill the gap. "Things went from very good to mediocre," he recalls. "Generally over time, things have kind of gone backwards." There were other problems as well. Some residents, "especially older people, especially if they came from another country," did not like the trees and objected to plantings in front of their homes. Some tree species gave tree-planting a bad name: "They made a lot of mistakes early on," admits Buikema. "Like those poplar trees that buckled the sidewalks, got into gutters, and crept under basement floors. It was really the wrong tree. It grows like a weed. Before the hybrids were developed, sugar maples were also problematic." Former City Arborist Rebecca Feldberg concurs: "Choosing appropriate species for urban settings is probably the biggest lesson learned. In the past, species were chosen that grew fast. This resulted in many trees being planted that were weak wooded and broke up in storms. . . . There was also little thought to whether exotic species . . . might become invasive." To make matters worse, many of the small trees city workers planted during Buikema's tenure suffered from vandalism and neglect after they were installed. Before long, his department was removing more trees than it was planting.[13]

Even in the face of budget cuts, labor shortages, and public resistance there is reason for hope. In Baltimore's Bolton Hill neighborhood, for instance, residents have embraced tree planting as an important community activity. Located northwest of downtown, Bolton Hill was once one of the city's most desirable neighborhoods. By the late 1950s, however, it was in sharp decline. Thanks to urban renewal funds, which placed an emphasis on street-tree planting, and the dedication and hard work of neighborhood tree advocates such as Ken Williams and George Lavdas, this neighborhood is now a shining example of what citizens can do to reforest Baltimore.[14]

Mike Galvin, former Supervisor of Urban and Community Forestry at the Maryland Department of Natural Resources, remains cautiously optimistic. He points to programs such as Yale University's Urban Resources Initiative (URI); the U.S. Department of Agriculture Forest Service's Revitalizing Baltimore program; the National Science Foundation-funded Baltimore Ecosystem Study (BES); and, most recently, TreeBaltimore. "They've all made a huge difference," Galvin insists. "Once you build the social and environmental equity—all these successes build on one another. It continues to build and build and build."[15]

Given the Yale School of Forestry's historical ties to Maryland—manifested in the careers and accomplishments of Fred Besley and Brooke Maxwell—one could argue that the Urban Resources Initiative was more of a next step than a new program. The roots of the URI can be traced to the late 1980s, when Ralph Jones, head of Recreation and Parks, challenged William (Bill) Burch, of the Yale School of Forestry, to conduct community-based work in Baltimore. At the time, Burch, who is the Hixon Professor of Natural Resource Management, was studying community-based resource management in Nepal. Unfortunately, Jones died before the two could reach a formal agreement. After attending Jones's funeral, Burch was more determined than ever to push ahead with the idea. Shortly thereafter, the Urban Resources Initiative—a partnership linking Yale, Baltimore, and the Parks and People Foundation—was born. Its creators envisioned using it as a way to build infrastructure in the community and manage the urban ecosystem. Parks and People would handle the administration; Yale would supply the students.

Starting with one intern, Morgan Grove in 1989, the program grew by leaps and bounds until there were seventeen interns working in the city in 1993. Soon, interns were diffusing information about community forestry to neighborhoods across the city, converting vacant lots, greening schoolyards, establishing pocket parks, and planting street trees. They worked to promote tree stewardship and community forestry. They collected survey data and compiled statistics. They produced the first brochure marketing the city's parks and floated the idea of a Gwynns Falls Trail. They began to involve local people and engage youth. The interns also established a good working relationship with City Arborist Jim Dicker and Park Superintendent Calvin Buikema. Collectively, the program interns brought "a beam of hope" to neglected parts of Baltimore. According to Burch, "The main focus was to develop these communities to use open space as a device to get people out of their houses and take charge of their neighborhoods."[16] "What these young people did was to bring a lot of new ideas," recalls Buikema. "They were young people. They were smart. Today, you can see the results of a lot of what they did."[17] Before long, Yale alumni in Baltimore were lending their support. Former Mayor Kurt Schmoke, a Yale graduate and member of the school's corporate board, endorsed the program, as did Martin O'Malley.[18]

Unfortunately, the effort began to lose steam. As Burch remembers it, Yale stopped putting money into the program and then momentum shifted to other projects. Without this critical resource base, the program withered away: "It's good to involve the local citizens adjacent to these trees, but they must be kept going. Yale interns organized the tree stewards and kept it going. . . . Unless you have tree stewards, . . . you haven't got a hope. If you don't have some endowment to keep it going and a group of young professionals to sustain the volunteers, . . . it collapses."[19] In the end, the greatest legacy of the URI may turn out to be the solid foundation it laid for other programs and projects, such as the long-term urban ecological research project known as the Baltimore Ecosystem Study.

Even when URI was flourishing, there were trouble spots in the city that fell through the cracks. Sometimes residents took matters into their own hands. With a laugh, Burch remembers the "urban guerillas" who tried surreptitiously to reforest parts of the city during the 1990s. The work of one "guerilla," now a planner involved with the TreeBaltimore project, was featured recently in the *Baltimore City Paper*:

Gary Letteron fingers the chaos of morning glory and pokeberry that obscure the beauty of small, potted trees. "I like weeds," says the urban-foresting force behind the Washington Village/Pigtown Neighborhood Planning Council. "There are already enough manicured places in the world." The Sowebo resident has cast his unconventional sights—seeing beauty where there is refuse, possibility where there is asphalt—on Southwest Baltimore's most neglected areas. . . . When he started planting trees throughout his neighborhood 15 years ago, though, he wasn't part of the establishment. He was a brigand street blaster, using money he earned from building sets for local theaters to rent jackhammers and drill into sidewalks, hauling away the rubble and planting trees in the 4-foot-deep holes. (For the same service, says Marion Bedingfield, a tree-service technician with the city's Forestry Division, independent contractors charge about $500 per tree-filled hole.) "It was a hoot," Letteron says of his urban-guerrilla days. "The police were after us because we planted hundreds of trees without permits. This was our version of graffiti."[20]

Fig. 6.2. Baltimore's legacy of planting trees continues at Fells Point in Baltimore.

Eventually, Letteron "ingratiated himself into the city's Forestry Division" and landed a job with Parks and People. Using live trees he hauls away from construction sites, Letteron and the volunteers he works with have planted hundreds of trees in Baltimore. "'There's nothing like coming upon a block full of concrete and leaving it with things growing there,' he says. . . . 'No question about it,' says Bedingfield, a frequent cohort on such efforts. 'He's the original community forester for Baltimore.'"[21]

Positive developments on the political front—"The last two mayors [Martin O'Malley and Sheila Dixon] have put green issues at the forefront of their agendas"—have convinced Galvin that Baltimore's goal of doubling its tree canopy in thirty years is not a pipe dream.[22] "I am optimistic that, if handled in partnership with the community, it can happen," affirms Jackie Carrera, Executive Director of Baltimore's Parks and People Foundation. "Mayor O'Malley set the goal; Mayor Dixon has embraced the goal."[23] To meet the challenge the city intends to protect existing trees, increase plantings on private lands, plant tree species best suited to the urban environment into which they are introduced, promote sustainable forestry practices on public and private land, design urban infrastructure to maximize tree-growing environments, foster interagency cooperation, and partner with public and private institutions.[24] The effort will also require enlisting the support of volunteers, drawing on new sources of revenue, and, most importantly, maintaining the initiative as a priority. Another indication that the tide may be turning in favor of the city's trees is that Baltimore recently hired an arborist: "There was nobody for a while there with the title of city forester or city arborist," remarked Galvin during a recent interview. "There was no one steering the bus. Having that . . . position re-established in Parks and Rec[reation] was critical."[25]

While these are certainly very positive developments, numerous obstacles remain. Sometimes the odds seem overwhelming. At present, Baltimore's tree canopy covers just over twenty percent of the city, far below the nationwide average of thirty-three percent for cities.[26] Right now, only 2,000 trees a year are being planted, which is fewer than the number of dead or dying currently being removed, according to Feldberg.[27] While both the mayor and the governor are supportive of Baltimore's efforts to expand its tree canopy, this has not always been the case. "Every few years you can get a new mayor," observes Buikema. "Each mayor comes in with a different agenda. It all comes down to different priorities and different governments."[28] If Baltimore hopes to reach its tree canopy goal, the program needs to be funded adequately and consistently. "Keeping the local funding stream going is important," notes Galvin.[29]

Guy Hager, Director of Great Parks, Clean Streams & Green Communities at Parks and People, identifies another critical challenge, one that a succession of city foresters over the years has struggled to address. "I think the most important thing is not the numeric goal but working systematically toward enhancement of tree survivability and better tree distribution throughout the city," he notes. "I would like to see the city grow canopy where it is not currently high (inner city neighborhoods) and have healthy trees in these communities. I see this as an environmental justice issue that would, if achieved, improve several quality of life issues in these neighborhoods."[30]

Even relatively simple issues, such as providing adequate soil and watering, can present challenges. "They've got to solve the watering problem," says Buikema. The city needs "to be able to keep the newly planted trees alive at least for the first couple of years until they can survive on their own. . . . They can't rely on the citizens; it doesn't seem to work."[31] Perhaps if more people like Ken Williams, George Lavdas, and Gary Letteron stepped up to meet the challenge, the initiative will succeed. Individuals such as Gene DeSantis provide hope for the future: "There were not many trees in California, where I grew up," recalls DeSantis. "It was a slum, a trashy kind of environment." As a teenager, he felt "worthless." That is, until he started planting trees. "Doing the trees made me feel like I was worth something," he acknowledges. "Now I'm doing it to help others. Planting trees provides a cleaner environment for children. The air's cleaner, you feel better. It's better than asphalt and trash." Since 1978, DeSantis has planted 12,401 trees in Baltimore and the number keeps growing.[32] The city will need more "tree stewards" like Williams, Lavdas, Letteron, and DeSantis, if it is to meet its ambitious tree canopy goals, because the success of TreeBaltimore will depend heavily on citizen involvement. As the city's draft Urban Forest Management Plan makes clear, the document "fulfills the planning part. The entire civic community will be needed for implementation."[33]

More than two decades ago, Anne Whiston Spirn pointed out that much of what comprises the urban forest today exists because we let it:

> Trees arching out over streets and green lawns with shade trees are a popular landscape aesthetic. To most people, nature in the city *is* trees, shrubs, and grass in streets, parks, and private yards, but these are actually the least "natural" of urban plant communities. The composition and arrangement of these plants are functions of land use and fashion rather than natural processes.[34]

Or, as Bill Burch likes to say, "Growing a tree in a city is a helluva tough job." The implication is clear. These aesthetic amenities come with a price tag attached. They "survive only with careful maintenance and decline with neglect." When "restricted municipal budgets" are added to the mix, an already "hostile city environment" becomes even more unfavorable, resulting in "disastrous consequences for this high-maintenance landscape."[35] As Henry W. Lawrence reminds us, much of the present-day urban forest was planted "in an age when labor costs were much lower and the destructive pressures much less" than they are today. "Large parks and public street trees in particular," he writes, "need a kind of regular care and maintenance that has become more and more difficult for some cities to finance in recent years."[36] Tree advocates in Baltimore know these lessons only too well.

The Next 100 Years

As Marylanders, in general, and government officials, in particular, deliberate over how best to manage the state's natural resources, they would do well to consider how these resources have been administered in the past. State forests and parks did not magically appear on the map; street trees and other "natural" features of the urban landscape did not miraculously spring from the ground. Rather, they are the outward manifestation of complex physical processes, social and political negotiations, land-use decisions, management strategies, and investments of time and money, concentrated and accumulated in specific locations. When these resources have been managed skillfully and with care, benefits have accrued to both the city and state. When their support has flagged or they have been otherwise neglected, these benefits have diminished or vanished altogether.

The problems that confront us in the twenty-first century are every bit as challenging—if not more so—than the ones Fred Besley and Brooke Maxwell faced a century ago. At the state level, disease, invasion of exotic species, and global warming are all serious threats. Even more worrisome, an insatiable desire to develop private timberlands is fragmenting today's forest. If this trend continues, the only large patches of forested land remaining will be isolated parcels owned by the state. Given the problems we face, it is not practicable to think we can rely solely on public lands to provide the services we need. More than sixty years after publication of Aldo Leopold's *A Sand County Almanac and Sketches Here and There* (1949) we are still searching for a land ethic to guide our resource management decisions: one that views a wooded acre as something more than a set of building blocks for yet another suburban subdivision.

In Baltimore, interest in urban forestry has ebbed and flowed over the years. Initially, concern over the poor condition of the city's trees sparked action on the part of local citizens groups. Soon a professional forester was hired and a Division of Forestry established. However, by the end of World War II, if not much earlier, the city was removing more trees than it was planting. Another infusion of funds was needed during the 1960s to reverse losses in forest canopy. Today, Baltimore's urban forest is again attracting attention in an effort to stem the tide of tree loss. If there is one lesson that can be drawn from past experience, it is this: we can no longer afford to invest time and money in such a haphazard way. Only a broad and sustained effort—one that enlists the cooperation and support of public and private partners—stands a chance of success.

While we all benefit from products derived from the forest, few of us have firsthand knowledge of the damage we inflict on these ecosystems as a result of our consumption. This is because our landscapes of consumption—our lumberyards, shopping malls, strip and housing developments, among others—are generally far removed from the places where forest resources are extracted. And while we all profit from the ecological services trees provide—shade, clean air, reduced runoff, and improved water quality—we too often take them for granted. Only when delivery of these services is impaired or eliminated do we lament their passing. As we look ahead to the next one hundred years, it is imperative that we recognize the impact that our resource management decisions, land-use practices, and lifestyle choices have on our forest resources. Only then can we begin to plan more wisely for the future. [37]

Epilogue

On Behalf of the Public Good

"My general health is excellent. Although sight is considerably limited and hearing not acute, I present considerable vigor for my years. Greetings and all good wishes for the Class of 1904."[1] So Fred Besley concluded his final entry for the October 1959 issue of the *Yale Forest School News*. Although the years were taking a toll on him physically, he reported that he was still "able to get in the woods, to look after and manage our timber holdings . . . in S. E. Maryland." He also hoped to organize a reunion of classmates in the near future.[2]

The following month, the *Baltimore Sun* announced the retirement of Brooke Maxwell. Although he agreed to serve as "associate landscape architect for the downtown greenery program," he turned down an offer to stay on as Director of Parks and Recreation. Within a year of mandatory retirement, he had been asked to remain on the job by the Board of Estimates. "But I turned it down," he conceded. "You want to have some fun before you get through. I just don't want the pressure of this job anymore."[3] Officially, Brooke Maxwell's last day was January 1, 1960. Later that year, on November 7, Fred Besley died of heart failure at the age of eighty-eight.[4] It was quite literally the end of an era with respect to forestry in both Baltimore and Maryland.

For their groundbreaking work in forestry and forest conservation, both Besley and Maxwell may truly be called pioneers. Besley's contributions on behalf of the state of Maryland during his thirty-six-year career—from fire fighting and education outreach to the establishment of a state nursery, to the creation of a network of state forests and parks—have accorded him legendary status in state forestry circles. Though less well known, Maxwell's efforts on behalf of Baltimore's trees and green spaces are no less remarkable, especially considering that much of the time he spent as a landscape architect in private industry was focused on city beautification. Of course, neither Besley nor Maxwell could have accomplished their deeds alone, but each must be credited for blazing a trail that others would follow.

To be sure, Besley had his critics. Some in the professional forestry community charged that he moved too slowly when it came to building the state forest system. Others claimed—rightfully so—that he favored the establishment of forests over parks. In the eyes of commercial nurserymen, his policies ran roughshod over their business interests. He could be stubborn and irascible, especially when he suspected that the welfare of the forestry program was at stake or that his authority was in some way being undermined. Yet no one could question Besley's dedication to the cause of scientific forest management or his loyalty to the Forestry Department.

In a tribute to Maxwell published after his death in 1969, the *Baltimore Sun* commented that his "greatest single virtue as the city's director of parks and recreation was his passionate belief in the importance of his job. Grass, shrubs and trees were not, to Mr. Maxwell, simply nice things to have in a city if you could get them. He considered them necessities and he fought at every turn for more." Choosing to "devote himself primarily to public life," the editorialist concluded that not just Baltimore, "but the whole state" were better off "for Mr. Maxwell's life-long devotion."[5]

Perhaps it was a product of their training at Yale, or maybe they were simply imbued with the spirit of the Progressive Era. Either way, Besley and Maxwell and the men and women with whom they collaborated and worked felt compelled to combine their forestry training with a call to public service. Following a Yale Forest School reunion during the 1920s, Besley commented that, although everyone in attendance "had a grand good time," he was most impressed by the "serious, sober thinking that was put into all of the discussions, and the substantial results" that were achieved. "The thought of public service that seemed to permeate the whole reunion," he marveled, "made each one feel that he was engaged in constructive work for public good. We all came back to our work with a feeling of pride in the profession, greater reverence for the school, and with a determination to make our work count."[6] Alongside street trees and state forests, this unflagging enthusiasm to serve the "public good" may prove to be our greatest inheritance from America's—and Maryland's— first generation of foresters.

Fig. A.1. James H. "Uncle Jim" Gambrill (center) was a leading conservationist in Maryland. He is pictured here with two unidentified men on Baumiller State Road and Purdy State Road at Roth Rock Tower, Garrett County.

Appendix A

Maryland State Forest Lands Acquired as of 17 May 1934

Swallow Falls

Date	From Whom Acquired	Subdivision Name	Gift Area	Purchase Area	Per Acre	Total	Total Acreage
1906	Robert and John Garrett	Skipnish	888.00				888.00
		Brier Ridge	823.00				823.00
		Kindness	206.00				206.00
1917	Henry and Julian LeRoy White	Herrington Manor	656.00				656.00
1918	John E. Wilson	Wilson Purchase		57.00	10.00	570.00	57.00
1923	Jonas and Minnie Sines	False Alarm		168.30	10.00	1683.00	168.30
1928	Groves Estate	Groves Tract		27.00	10.00	270.00	27.00
1930	Kendall, Sheriff's Sale	Piney Run Lots		1,142.50	1.25	1,503.75	1,142.50
1931	Offutt Estate	Offutt Lots		568.00	3.00	1,704.00	568.00
1932	Tax Deed, Mineral Rights	Herrington Manor				273.98	
						6,004.73	4,535.80

Appendix A (*continued*)

Fort Frederick

Date	From Whom Acquired	Subdivision Name	Gift Area	Purchase Area	Per Acre	Total	Total Acreage
1922	Homer J. Cavanaugh and wife			189.00	61.00	11,500.00	189

Patapsco

Date	From Whom Acquired	Subdivision Name	Gift Area	Purchase Area	Per Acre	Total	Total Acreage
1913	John Glenn, Jr.	Glenn Tract	40.15				40.15
1913	Wm. Asshton	Asshton Tract		40.85		2,330.38	40.85
1913	Stephen Harwood	Harwood Tract		101.80		5,599.00	101.80
1913	Howard Mann	Mann Tract		103.22		3,096.60	103.22
1913	Edward R. Dennis	Dennis Tract		15.40		462.00	15.40
1913	Hanson Bros.	Hanson Tract		26.30		1,183.50	26.30
1913	Elliott	Sucker Branch		5.00		225.00	5.00
1914	R. C., J. P., and R. Norris	Norris Tract	34.90				34.90
1914	Avalon Realty Co.	Avalon Tract		191.10		11,466.00	191.10
1914	Roy R. Clark	Clark Tract		30.01		900.00	30.01
1915	Wm. L. Glenn	Glenn Tract	56.00				56.00
1915	Wm. L. Glenn	Glenn Tract	56.00				56.00

Appendix A (*continued*)

Patapsco *(continued)*

Date	From Whom Acquired	Subdivision Name	Gift Area	Purchase Area	Per Acre	Total	Total Acreage
1915	Wm. L. Glenn	Glenn Tract	33.42				33.42
1915	Hanson Estate	Hanson Tract		129.00		6,450.00	129.00
1917	Martha A. Isaacs	Warden's House		8.25		1,500.00	8.25
1917	Mangers Children's Home	Mangers Tract		22.00		1,100.00	22.00
1918	Theo. Marburg	Marburg Tract		73.54		4,412.40	73.54
1920	C. A. Gambrill Co.	Right-of-Way	.05				.05
1920	Clinton L. Riggs	Riggs Tract		15.30		918.00	15.30
1926	W. U. Dickey and Sons	Dickey		190.00		10,000.00	190.00
						4,9642.28	1,116.29

Savage River

Date	From Whom Acquired	Subdivision Name	Gift Area	Purchase Area	Per Acre	Total	Total Acreage
1929	Bond Fish and Game Association	Bond Tract		5,023.00	2.00	10,046.00	5,023.00
1930	Bond Fish and Game Association	Bond Tract		4,536.00	2.00	9,072.00	4,536.00
1930	John Dimeling	Dimeling and Bloom		6,619.00	2.00	13,238.00	6,619.00
1932	Virgil Sines			35.00	2.00	70.00	35.00

Appendix A (*continued*)

Savage River (*continued*)

Date	From Whom Acquired	Subdivision Name	Gift Area	Purchase Area	Per Acre	Total	Total Acreage
1932	Harvey Broadwater			40.30	2.00		40.30
				24.00	2.00		24.00
				26.00	2.00	180.60	26.00
				24.00	2.00		24.00
				26.00	2.00	180.60	26.00
1932	Nelson Broadwater			25.50	2.00	51.00	25.50
1932	McAndrew Brothers	Mill Site		86.00		2,750.00	86.00
1932	Floyd Broadwater	Mill Site		2.00		25.00	2.00
						35,432.60	16,416.80

Seth Demonstration

Date	From Whom Acquired	Subdivision Name	Gift Area	Purchase Area	Per Acre	Total	Total Acreage
1929	Mary W. Seth	Seth Demonstration	65.00				65.00

Appendix A (*continued*)

Doncaster

Date	From Whom Acquired	Subdivision Name	Gift Area	Purchase Area	Per Acre	Total	Total Acreage
1930	Henrico Lumber Co.			1,178.66	4.00	4808.93	1,156.78
1932	Walter Mitchell	Mitchell		100.00	4.00	400.00	100.00
						5,208.93	1,256.78

Cedarville

Date	From Whom Acquired	Subdivision Name	Gift Area	Purchase Area	Per Acre	Total	Total Acreage
1930	Beta Realty Co.	National Stock Farm		2295.30			
		Little Philadelphia		64.50			
		Edward Mudd		217.20	4.00	10,524.00	2,631.00
1932	Miles Estate	Miles Tract		100.00	3.00	300.00	100.00
						10,824.00	2,731.00

Appendix A (*continued*)

Potomac

Date	From Whom Acquired	Subdivision Name	Gift Area	Purchase Area	Per Acre	Total	Total Acreage
1931	Manor Mining and Manufacturing Co.	Manor Tract		5,979.00	.50	3,014.50	5,979.00
1931	Elizabeth R. P. Hubbard	Hubbard Tract Lot # 404		43.00	1.16	50.00	43.00
1933	Robert Sheckells	Sheckells Tract		400.00		175.00	400.00
1934	Jones and Sheckells	Jones and Sheckells		620.00	1.25	775.00	620.00
						4,014.50	7,042.00

Green Ridge

Date	From Whom Acquired	Subdivision Name	Gift Area	Purchase Area	Per Acre	Total	Total Acreage
1931	Grove and George Heirs	George Tract		1,737.22	3.00	5,211.66	1,737.22
1931	Queen E. Potts	Potts Tract		235.00			
				40.00	3.00	825.00	275.00
1931	Jas. B. Price	Price Tract		394.25	3.00	1,182.75	394.25
1932	H. A. Carpenter	Carpenter Tract		1,980.83	3.00	5,942.49	1,980.83
1932	Allegany Orchard Co.	Orchard Tract		11,777.61	1.75	19,250.00	11,777.61
1932	Henry F. Harrison	Harrison Tract		10.00	1.75	17.50	10.00

Appendix A (*continued*)

Green Ridge (*continued*)

Date	From Whom Acquired	Subdivision Name	Gift Area	Purchase Area	Per Acre	Total	Total Acreage
1933	Margaret G. Price	Price Tract		340.00	1.50	510.00	340.00
1933	Queen E. Potts	Potts Tract		72.00		175.	72.00
						33,114.40	16,586.91

Pocomoke

Date	From Whom Acquired	Subdivision Name	Gift Area	Purchase Area	Per Acre	Total	Total Acreage
1932	Frank E. Hudson	Hudson Tract		740.00		4,758.20	740.00
1934	C. F. Chandler	Chandler Tract		453.00		3,000.00	453.00
						7,758.20	1,193.00

Washington Monument

Date	From Whom Acquired	Subdivision Name	Gift Area	Purchase Area	Per Acre	Total	Total Acreage
1933	Historical Society et al., Washington County	Washington Monument	11.00				11.00

Source: Records of the President's Office, Series VII, Box 17, State Forests Acquired as of May 17, 1934, University of Maryland, University Archives, College Park, Maryland.

Fig. A.2. "Firefighting equipment," ca. 1950s. In some ways, little had changed in fire-fighting equipment since Besley's day. Photograph by M. E. Warren.

Appendix B

Trees Planted and Removed in Baltimore, 1957–1965

Year	Trees Planted		Trees Removed	
1957	Parks	1,349	Parks	407
	Highways	2,008	Highways	1,004
	Total	3,357	Total	1,411
1958	Parks	128	Parks	822
	Highways	379	Highways	790
	Total	507	Total	1,612
1959	Parks	249	Parks	533
	Highways	2,048	Highways	651
	Total	2,297	Total	1,184
1960	Parks	182	Parks	453
	Highways	1,061	Highways	716
	Other	1,000		
	Total	2,243	Total	1,169

Appendix B (*continued*)

Year	Trees Planted		Trees Removed	
1961	Parks	133	Parks	706
	Highways	972	Highways	1,521
	Other	171		
	Total	1,276	Total	2,227
1962	Parks	408	Parks	691
	Highways	831	Highways	873
	Other	387	Total	1,564
	Total	1,626	(plus 1,097 by contract)	
1963	Parks	265	Parks	964
	Highways	1,148	Highways	913
	Other	205	Total	1,877
	Total	1,618	(plus 1,479 by contract)	
1964	Parks	429	Parks	375
	Highways	1,101	Highways	654
	Other	144	Total	1,029
	Total	1,674	(plus 1,223 by contract)	
1965	Parks	448	Parks	560
	Highways	6,208	Highways	715
	Other	23	Total	1,275
	Total	6,679	(plus 685 by contract)	

Source: Reports of the Park Commission to the Mayor and City Council of Baltimore (1957–1965).

Notes

A Note to the Reader: Page numbers are unavailable for those sources containing information found in newspaper clippings from scrapbooks. Also, the epigraph on page iv is from William Blake's "Auguries of Innocence" in David V. Erdman, ed., *The Complete Poetry and Prose of William Blake* (Berkeley: University of California Press, 1982), 490.

Introduction

1. Samuel P. Hays chronicles the conflicts that have pitted proponents of "commodity forestry" against promoters of "ecological forestry" in *Wars in the Woods: The Rise of Ecological Forestry in America* (Pittsburgh: University of Pittsburgh Press, 2007).

2. Joe Palazzolo, "City plans for greener future," *Baltimore Sun* (30 March 2006); Michael F. Galvin, J. Morgan Grove, and Jarlath O'Neil-Dunne, *A Report on Baltimore City's present and potential Urban Tree Canopy*, prepared for the Honorable Martin O'Malley, Mayor, City of Baltimore. Annapolis, Maryland Department of Natural Resources, January 19, 2006.

3. Campbell Gibson, *Population of the 100 Largest Cities and Other Urban Places in the United States: 1790 to 1990*. Population Division Working Paper No. 27. Washington, DC: U.S. Bureau of the Census, 1998).

4. In his book, *City Trees: A Historical Geography from the Renaissance through the Nineteenth Century* (Charlottesville: University of Virginia Press, in association with the Center for American Places, 2006), Henry W. Lawrence writes: "It was not until the end of the nineteenth century . . . that American city governments began to take control of street tree planting. In New York, for instance, it was only in 1899 that an act of the state legislature gave cities jurisdiction over street trees and 1902 that New York City delegated this authority to the Park Board of Greater New York," (247–48).

5.   In 1958, Baltimore Mayor Thomas D'Alesandro stated: "It would be nice if Baltimore could become known as the city of a million trees" ("Mayor Wants Tree Program," *Baltimore Evening Sun*, 12 June 1958); and Heather Dewar. "Study Shows a Shrinking Forest," *Baltimore Sun* (15 December 2003). According to this article, there are an estimated 2.6 million trees comprising Baltimore's tree canopy. The Division of Forestry manages approximately 500,000 trees.

6.   There are notable exceptions, including William H. Rivers, "Massachusetts State Forestry Programs," in Charles H. W. Foster, ed., *Stepping Back to Look Forward: A History of the Massachusetts Forests* (Cambridge: Harvard Forest Press, 1998), 147–219; and E. Gregory McPherson and Nina Luttinger, "From Nature to Nurture: The History of Sacramento's Urban Forest," *Journal of Arboriculture*, Vol. 24, No. 2 (1998): 72–88.

7.   "Those Among Us You Should Know," *American Forests*, Vol. 37, No. 3 (March 1931): 179.

Chapter One

The chapter epigraphs come from Gifford Pinchot, *Breaking New Ground* (New York: Harcourt, Brace and Company, 1947), 23, and George B. Sudworth, "The Forests of Allegany County," in *Maryland Geological Survey: Allegany County* (Baltimore: The Johns Hopkins University Press, 1990), 279.

1.   Keith D. Wiebe, Ababayehu Tegene, and Betsey Kuhn, "Land Tenure, Land Policy, and the Property Rights Debate," in Harvey M. Jacobs, ed., *Who Owns America?: Social Conflict Over Property Rights* (Madison: University of Wisconsin Press, 1998), 79–93.

2.   Carolyn Merchant, *The Columbia Guide to American Environmental History* (New York: Columbia University Press, 2002), 123.

3.   Ibid., 124–25.

4.   Michael Williams, "Clearing the United States' forests: pivotal years 1810–1860," *Journal of Historical Geography*, Vol. 8, No. 1 (1982): 14–15.

5.   Douglas W. MacCleery, *American Forests: A History of Resiliency and Recovery* (Durham, NC: U.S. Department of Agriculture Forest Service, in cooperation with the Forest History Society, 1993).

6. Michael Williams, *Americans & Their Forests: A Historical Geography* (Cambridge: Cambridge University Press, 1989).

7. MacCleery, *American Forests*.

8. Wiebe, "Land Tenure"; and Williams, *Americans & Their Forests*.

9. Howard Zinn, *A People's History of the United States: 1492–Present* (New York: Perennial Classics, 2003).

10. James Chace, 1912: *Wilson, Roosevelt, Taft & Debs—The Election That Changed the Country* (New York: Simon & Schuster, 2004), 240.

11. Neil M. Maher, "'A Conflux of Desire and Need': Trees, Boy Scouts, and the Roots of Franklin Roosevelt's Civilian Conservation Corps," in Henry L. Henderson and David B. Woolner, eds., *FDR and the Environment* (New York: Palgrave MacMillan, 2005), 57.

12. Samuel P. Hays, *Conservation and the Gospel of Efficiency: The Progressive Conservation Movement, 1890–1920* (New York: Atheneum, 1969), xii–xiii, 29, and 32.

13. Donald J. Pisani, "Forests and Conservation, 1865–1890," in Char Miller, ed., *American Forests: Nature, Culture, and Politics* (Lawrence: University Press of Kansas, 1997), 15.

14. Robert W. Miller, "The History of Trees in the City," in Gary Moll and Sara Ebenreck, eds., *Shading Our Cities: A Resource Guide for Urban and Community Forests* (Washington, DC.: Island Press, 1989), 32–34; and Blake Gumprecht, "Transforming the Prairie: Early Tree Planting in an Oklahoma Town," *Historical Geography*, Vol. 29 (2001): 112–34.

15. Miller, "History of Trees in the City."

16. Harold K. Steen, "The Beginning of the National Forest System," in Char Miller, ed., *American Forests: Nature, Culture, and Politics* (Lawrence: University Press of Kansas, 1997), 49–68.

17. Williams, *Americans & Their Forests*, 412–18.

18. Ibid., 442.

19. Hal K. Rothman, "'A Regular Ding-Dong Fight': The Dynamics of Park Service-Forest Service Controversy During the 1920s and 1930s," in Char Miller, ed., *American Forests: Nature, Culture, and Politics* (Lawrence: University Press of Kansas, 1997), 109 and 112; and Gifford Pinchot, Forester, U.S. Department of Agriculture Forest Service, *The Use Book: Regulations and Instructions for the Use of the National Forest Reserves*, 1906.

20. Mark Baker, *Community Forestry in the United States: Learning from the Past, Crafting the Future* (Washington, DC: Island Press, 2003), 37.

21. Christopher F. Meindl, Derek H. Alderman, and Peter Waylen, "On the Importance of Environmental Claims-Making: The Role of James O. Wright in Promoting the Drainage of Florida's Everglades in the Early Twentieth Century," *Annals of the Association of American Geographers*, Vol. 92, No. 4 (2002): 682; see, also, Craig E. Colten, "Industrial Topography, Groundwater, and the Contours of Environmental Knowledge," *The Geographical Review*, Vol. 88, No. 2 (1998): 199–218.

22. Christopher G. Boone and Ali Modarres, *City and Environment* (Philadelphia: Temple University Press, 2006), 160.

23. Jon Peterson, "The Evolution of Public Open Space in American Cities," *Journal of Urban History*, Vol. 12, No. 1 (1985): 75–76.

24. For an examination of how cemeteries were utilized to meet the recreational needs of urban residents in the years before large-scale park construction, see Thomas Bender, "The 'Rural' Cemetery Movement: Urban Travail and the Appeal of Nature," in Neil Larry Shumsky, ed., *The Physical City: Public Space and the Infrastructure* (New York: Garland Publishing, Inc., 1996), 2–17.

25. Henry W. Lawrence, "Changing Forms and Persistent Values: Historical Perspectives on the Urban Forest," in Gordon A. Bradley, ed., *Urban Forest Landscapes: Integrating Multidisciplinary Perspectives* (Seattle: University of Washington Press, 1995), 17; and D. B. Botkin and C. E. Beveridge, "Cities as Environments," *Urban Ecosystems*, Vol. 1, No. 1 (1997): 3–19.

26. Lawrence, ibid., 19–20.

27. Furman Lloyd Mulford, "Street and Roadside Trees," *American Forests and Forest Life*, Vol. 33, No. 403 (July 1927): 404.

28. Miller, "History of Trees in the City"; Lawrence, "Changing Forms and Persistent Values"; and Botkin and Beveridge, "Cities as Environments."

29. Julie Tuason, "*Rus in Urbe*: The Spatial Evolution of Urban Parks in the United States, 1850–1920," *Historical Geography*, Vol. 25, (1997): 124–47; Roy Rosenzweig and Elizabeth Blackmar, *The Park and the People: A History of Central Park* (Ithaca: Cornell University Press, 1992); and Witold Rybczynski, "Why We Need Olmsted Again," *Wilson Quarterly*, Vol. 23, No. 3 (1999): 15–22. See, also, David Schuyler, *The New Urban*

Landscape: The Redefinition of City Form in Nineteenth-Century America (Baltimore: The Johns Hopkins University Press, 1986).

    30.   Terence Young, "Modern Urban Parks," *The Geographical Review*, Vol. 85, No. 4 (1995): 537.

    31.   Lawrence, "Changing Forms and Persistent Values," 26–27; Virginia I. Lohr et al., "How Urban Residents Rate and Rank the Benefits and Problems Associated with Trees in Cities," *Journal of Arboriculture*, Vol. 30, No. 1 (2004): 28–35; Scott E. Maco and E. Gregory McPherson, "Assessing Canopy Cover Over Streets and Sidewalks in Street Tree Populations," *Journal of Arboriculture*, Vol. 28, No. 6 (2002): 270–76; John Dwyer, Herbert Schroeder, and Paul Gobster, "The Deep Significance of Urban Trees and Forests," in R. H. Platt et al., eds., *The Ecological City: Preserving and Restoring Urban Biodiversity* (Amherst: University of Massachusetts Press, 1994), 137–50; Stephen Kaplan, "The Urban Forest as a Source of Psychological Well-Being," in Gordon A. Bradley, ed., *Urban Forest Landscapes: Integrating Multidisciplinary Perspectives* (Seattle: University of Washington Press, 1995), 101–08; William E. Hammitt, "Urban Forests and Parks as Privacy Refuges," *Journal of Arboriculture*, Vol. 28, No. 1 (2002): 19–26; Kathleen L. Wolf, "Freeway Roadside Management: The Urban Forest Beyond the White Line," *Journal of Arboriculture*, Vol. 29, No. 3 (2003): 127–36; and Steven J. Anlian, "Building a Sense of Place," in *Shading Our Cities: A Resource Guide for Urban and Community Forests* (Washington DC.: Island Press; 1989), 106–11. Summit and McPherson report that shade and aesthetics play "a very large role" in a private landowner's decision to plant a tree. See Joshua Summit and E. Gregory McPherson, "Residential Tree Planting and Care: A Study of Attitudes and Behavior in Sacramento, California," *Journal of Arboriculture*, Vol. 24, No. 2 (1998): 89–96.

    32.   E. Gregory McPherson, "Urban Forestry Issues in North America and Their Global Linkages." Prepared for the 20th Session of the North American Forestry Commission, Food and Agriculture Organization of the United Nations, St. Andrews, New Brunswick, Canada, June 12–16, 2000, 1–7; Timothy Beatley, *Green Urbanism: Learning from European Cities* (Washington, DC.: Island Press, 2000); Sara Ebenreck, "The Values of Trees," in *Shading Our Cities: A Resource Guide for Urban and Community Forests* (Washington, DC: Island Press; 1989), 49–57: and David J. Nowak, Daniel E. Crane, and John F. Dwyer, "Compensatory Value of Urban Trees in the United States," *Journal of Arboriculture*, Vol. 28, No. 4 (2002): 194–99.

33.   Frances E. Kuo, "The Role of Arboriculture in a Healthy Social Ecology," *Journal of Arboriculture*, Vol. 29, No. 3 (2003): 148–55; Lynne M. Westphal, "Urban Greening and Social Benefits: A Study of Empowerment Outcomes," *Journal of Arboriculture*, Vol. 29, No. 3 (2003): 137–47; Kathleen L. Wolf, "Public Response to the Urban Forest in Inner-City Business Districts," *Journal of Arboriculture*, Vol. 29, No. 3 (2003): 117–26; Robert J. Laverne and Kimberly Winson-Geideman, "The Influence of Trees and Landscaping on Rental Rates at Office Buildings," *Journal of Arboriculture* Vol. 29, No. 5 (2003): 281–90; Jim Simpson and Greg McPherson, "Tree Planting to Optimize Energy and CO_2 Benefits," *Proceedings from the National Urban Forest Conference*, September 5-7, Washington, DC, 1–3; E. Gregory McPherson, Rowan A. Rowntree, and J. Alan Wagar, "Energy-Efficient Landscapes," in Gordon A. Bradley, ed., *Urban Forest Landscapes: Integrating Multidisciplinary Perspectives* (Seattle: University of Washington Press, 1995), 150–60; and Stephanie Pincetl and Elizabeth Gearin, "The Reinvention of Public Green Space," *Urban Geography*, Vol. 26, No. 5 (2005): 365–84.

34.   Anne Whiston Spirn, *The Granite Garden: Urban Nature and Human Design* (New York: Basic Books, 1984), 175.

35.   Michael F. Galvin, "A Methodology for Assessing and Managing Biodiversity in Street Tree Populations: A Case Study," *Journal of Arboriculture*, Vol. 25, No. 3 (1999): 124–28; and Martin F. Quigley, "Street Trees and Rural Conspecifics: Will Long-Lived Trees Reach Full Size in Urban Conditions?" *Urban Ecosystems*, Vol. 7, No. 1 (2004): 29–39.

36.   William F. Elmendorf, Vincent J. Cotrone, and Joseph T. Mullen, "Trends in Urban Forestry Practices, Programs, and Sustainability: Contrasting a Pennsylvania, U.S., Study," *Journal of Arboriculture*, Vol. 29, No. 4 (2003): 237–47; Gary Moll, "Improving the Health of the Urban Forest," in Gary Moll and Sara Ebenreck, eds., *Shading Our Cities: A Resource Guide for Urban and Community Forests* (Washington DC.: Island Press; 1989), 119–29; Joseph Poracsky and Mark Scott, "Industrial-Area Street Trees in Portland, Oregon," *Journal of Arboriculture*, Vol. 25, No. 1 (1999): 9–17; John F. Dwyer, David J. Nowak, Mary Heather Noble, and Susan M. Sisinni, *Connecting People with Ecosystems in the Twenty-First Century: An Assessment of Our Nation's Urban Forests*. A Technical Document Supporting the 2000 USDA Forest Service RPA Assessment (Portland, OR: USDA Forest Service, 2000); John F. Dwyer, David J. Nowak, and Mary Heather Noble, "Sustaining Urban Forests," *Journal of Arboriculture*,

Vol. 29, No. 1 (2003): 49–55; and David Despot and Henry Gerhold, "Preserving Trees in Construction Projects: Identifying Incentives and Barriers," *Journal of Arboriculture*, Vol. 29, No. 5 (2003): 267–80.

37.   Boone and Modarres, *City and Environment*, xi; and Beatley, *Green Urbanism*, 197.

38.   Gary Moll, "Urban Forestry: A National Initiative," in Gordon A. Bradley, ed., *Urban Forest Landscapes: Integrating Multidisciplinary Perspectives* (Seattle: University of Washington Press, 1995), 12.

39.   Father Andrew White, "An Account of the Colony of the Lord Baron of Baltamore, 1633," in C. C. Hall, ed., *Narratives of Early Maryland, 1633–1684* (New York: Charles Scribner's Sons, 1910), 8–9.

40.   Benjamin Silliman, *Extracts from a Report Made to the Maryland Mining Company on the Estate of Said Company, in the County of Allegany, Maryland* (New York: Scatcherd and Adams, 1838), 5.

41.   Raphael Semmes, *Baltimore: As Seen by Visitors, 1783–1860, Studies in Maryland History, No. 2* (Baltimore: Maryland Historical Society, 1953), 4, 60, 109, and 114.

42.   Jack Temple Kirby, *Poquosin: A Study of Rural Landscape and Society* (Chapel Hill: The University of North Carolina Press, 1995), 219.

43.   William Bullock Clark, *Maryland Geological Survey: Allegany County* (Baltimore: The Johns Hopkins [University] Press, 1900); and William Bullock Clark, *Report on the Physical Features of Maryland: Together with an Account of the Exhibits of Maryland Mineral Resources Made by the Maryland Geological Survey* (Baltimore: The Johns Hopkins [University] Press, 1906), 247–48.

44.   George B. Sudworth, "The Forests of Allegany County," in *Maryland Geological Survey: Allegany County* (Baltimore: The Johns Hopkins [University] Press, 1900), 279.

45.   *44th and 45th Annual Reports of the Board of Park Commissioners to the Mayor and City Council of Baltimore, For the Fiscal Years Ending December 31, 1903, 1904* (Baltimore: Wm. J. C. Dulany Company, City Printers, 1905), 34.

Chapter Two

The chapter epigraphs come from an address by President Theodore Roosevelt to the Society of American Foresters on March 26, 1903, as originally published in *Proceedings of the Society of American Foresters*, Vol. 1 (May 1905): 3–9 and reprinted in *Journal of*

Forestry, Vol. 98 (November 2000): 4; and Ralph R. Widner, "Maryland: Politics Do Not Fight Fires." In Ralph R. Widner, ed., *Forests and Forestry in the American States: A Reference Anthology* (Washington, DC: The National Association of State Foresters, 1968), 152.

1.   W. B. Greeley, "Future Trends in National and State Forestry," *American Forests and Forest Life*, Vol. 33, No. 397 (January 1927): 2–8.

2.   Ralph R. Widner, ed., *Forests and Forestry in the American States: A Reference Anthology* (Washington, DC: The National Association of State Foresters, 1968), xix–xx.

3.   Edna Warren, "Forests and Parks in the Old Line State," *American Forests*, Vol. 62, No. 10 (October 1956): 18.

4.   Warren, "Forests and Parks," 13–25 and 56–77; and Jack Temple Kirby, *Poquosin: A Study of Rural Landscape and Society* (Chapel Hill: The University of North Carolina Press, 1995), 219–21.

5.   *History of the Maryland Park Service* (http://dnr.maryland.gov/publiclands/100years/history.asp. Retrieved 26 November 2009.

6.   "First State Forester Fred W. Besley Honored at Golden Anniversary Celebration," *The Old Line Acorn* Vol. 13, No. 1 (March 1956): 3; and Widner, *Forests and Forestry*, 92–3.

7.   Robert and John W. Garrett, April 10, 1907, Garrett Bequest to the State of Maryland. Liber 54, folio 425, Land Records of Garrett County.

8.   F. W. Besley, "Progress in Maryland," *Yale Forest School News*, Vol. 1, No. 4 (October 1913); and Warren, "Forests and Parks," 15.

9.   "Fred W. Besley," Biographical Record of the Graduates and Former Students of the Yale Forest School. With Introductory Papers on Yale in the Forestry Movement and the History of the Yale Forest School (New Haven: Yale Forest School, 1913), 70; and Wells A. Sherman, "Typical Americans," January 31, 1936. This account of the life of Fred and Bertha Besley was written at the time of Bertha's death by Fred Besley's brother-in-law. In 1981, Besley's son, Lowell, reviewed the document and inserted names, places, and dates omitted in the original draft. According to Lowell, "The Besleys and the Shermans grew up together on adjacent farms in Virginia and Wells Sherman married Father's sister Elsie. Consequently, Uncle Wells had first hand knowledge of Father's youth and of his courtship of Mother as well as much information on earlier days."

10.   Sherman, "Typical Americans," 1.

11. Ibid., 2.

12. "Fred W. Besley," *Biographical Record*, 70; and "Fred Wilson Besley" (Jane Overington Wright Collection). This is an annotated timeline of major events in the life of Fred W. Besley.

13. Fred W. Besley, ca. 1956. "Partial Biography of Fred Wilson Besley" (Kirk Rodgers Collection), 1.

14. Besley, "Partial Biography"; "Bertha Adeline Simonds Besley" (Jane Overington Wright Collection). This is an annotated timeline of major events in the life of Bertha Adeline Simonds Besley; and Helen Besley Overington Interview. Interviewed by Jane Overington Wright, 14 March 1995 (Jane Overington Wright Collection).

15. "Bertha Adeline Simonds Besley"; and Helen Besley Overington Interview.

16. "Bertha Adeline Simonds Besley"; and Helen Besley Overington Interview; Sherman, "Typical Americans," 2.

17. Harold K. Steen, ed., *The Conservation Diaries of Gifford Pinchot* (Durham, NC: The Forest History Society, 2001), 89.

18. Will Barker, "Maryland's First State Forester," *American Forests*, Vol. 62, No. 10 (October 1956): 38.

19. Barker, "Maryland's First State Forester," 38 and 77–78.

20. Besley, "Partial Biography," 1.

21. Barker, "Maryland's First State Forester," 78.

22. Ibid.

23. Ibid.

24. Besley, "Partial Biography," 1.

25. Steen, *Conservation Diaries of Gifford Pinchot*, 112; and Gifford Pinchot, *Breaking New Ground* (New York: Harcourt, Brace and Company, 1947), 65.

26. Sherman, "Typical Americans," 3.

27. Besley, "Partial Biography," 1–2.

28. Besley, "Progress in Maryland"; Besley, "Partial Biography," 2.

29. Ibid.

30. Ibid.

31. Ibid.

32. "First State Forester," 4.

33.   Besley, "Partial Biography," 3.

34.   "Rites Held for Besley, First State Forester," *Baltimore Evening Sun* (10 November 1960).

35.   Besley, "Partial Biography," 3; "Fred W. Besley," *Biographical Record*, 71.

36.   Maryland Legislature, Forest Laws of Maryland, Acts of 1906, chapter 294, "An Act to establish a State Board of Forestry and to promote forest interests and arboriculture in the State," as amended in Chapter 161, Acts 1910, and Chapter 823, Acts of 1914.

37.   Besley, "Partial Biography," 3.

38.   Besley, "Progress in Maryland."

39.   Besley, "Partial Biography," 3; and Besley, "Progress in Maryland."

40.   Besley, ibid.; Barker, "Maryland's First State Forester," 38; Warren, "Forests and Parks," 19; and F. W. Besley, *The Forests of Maryland* (Baltimore: Maryland State Board of Forestry, 1916), 41.

41.   Besley, ibid.

42.   Gifford Pinchot, *The Training of a Forester*, 4th ed. (Philadelphia: J. B. Lippincott Company, 1937), 63. For a review of Pinchot's public relations efforts, see Stephen Ponder, "Federal News Management in the Progressive Era: Gifford Pinchot and the Conservation Crusade," *Journalism History*, Vol. 13, No. 1 (January 1986): 42–48; and Char Miller, "Old Growth: A Reconstruction of Gifford Pinchot's *Training of a Forester*, 1914–1937," *Forest & Conservation History*, Vol. 38, No. 1 (January 1994): 7–15. Austin Hawes provides a state-level perspective in "Connecticut: Scattering the Seeds," in Ralph R. Widner, ed., *Forests and Forestry in the American States: A Reference Anthology* (Washington, DC: National Association of State Foresters, 1968), 103.

43.   Fred W. Besley, *Maryland's Forest Resources: A Preliminary Report, Forestry Leaflet No. 7.* (Baltimore: Maryland State Board of Forestry, 1909), 3.

44.   Besley, "Partial Biography," 4.

45.   According to Besley's "Partial Biography," 4, "Special reports, with large scale forest maps in color were published for Allegany, Anne Arundel, Baltimore, Frederick, Garrett, Kent, Prince George's and Washington Counties upon completion of the work in these counties."

46.   F. W. Besley, "The Forests and Their Products," in *The Plant Life of Maryland* (Baltimore: The Maryland Weather Service, 1910), 379.

47.   Ibid.

48. Besley, *Maryland's Forest Resources*; F. W. Besley, *The Forests of Allegany County* (Baltimore: Maryland State Board of Forestry, 1912); and Besley, "Forests and Their Products," 364; and Besley, *Forests of Maryland*, 47–48.

49. Fred W. Besley, *The Wood-Using Industries of Maryland* (Baltimore: Maryland State Board of Forestry, 1919).

50. Besley, Forests and Their Products, 363 and 376; C. D. Mell, "The Forests of St. Mary's County," in *Maryland Geological Survey: St. Mary's County* (Baltimore: The Johns Hopkins [University] Press, 1907), 184; and F. W. Besley, "The Forests of Queen Anne's County," in *Maryland Geological Survey: Queen Anne's County* (Baltimore: The Johns Hopkins [University] Press, 1926), 162; and F. W. Besley, *The Forests of Garrett County* (Baltimore: The Maryland State Board of Forestry, 1914), 7–8.

51. Timothy Cochrane, "Trial By Fire: Early Forest Service Rangers' Fire Stories," *Forest & Conservation History*, Vol. 35, No. 1 (January 1991): 16–23; *Baltimore Sun*, 1, 2, and 16 April 1920; 6 May 1920; 9, 16, 26 January 1921;, 5, 6, and 19 February 1921; 18 May 1921; *Baltimore News*, 16 April 1920; 3 May 1920; 5 March 1921; and *Oakland Republican*, 19 May 1921.

52. William Bullock Clark, Report on the Physical Features of Maryland: Together with an Account of the Exhibits of Maryland Mineral Resources Made by the Maryland Geological Survey (Baltimore: The Johns Hopkins [University] Press, 1906), 248.

53. F. W. Besley, *The Forests of Anne Arundel County* (Baltimore: Maryland State Board of Forestry, 1915), 22.

54. Besley, "Partial Biography," 4.

55. Ibid.

56. "F. W. Besley," *Yale Forest School News*, Vol. 3, No. 4 (October 1915): 51.

57. Besley, "Partial Biography," 5.

58. Ibid.

59. F. W. Besley, *Report of the State Board of Forestry for 1908 and 1909* (Baltimore: The Johns Hopkins [University] Press, 1909), 7.

60. F.W. Besley, *Report of the State Board of Forestry for 1922 and 1923* (Baltimore: The Johns Hopkins [University] Press, 1923), 37.

61. Beatrice Ward Nelson, *State Recreation: Parks, Forests and Game Preserves* (Washington, DC: National Conference on State Parks, Inc., 1928), 111.

62. Anne Buckelew Cumming et al., "Forest Health Monitoring Protocol Applied to Roadside Trees in Maryland," *Journal of Arboriculture*, Vol. 27, No. 3, (2001): 127.

63. Maryland Legislature, Forest Laws of Maryland, Acts of 1914, Chapter 824, as amended in Chapter 548, Acts of 1916.

64. Tunis J. Lyon, "Maryland's Roadside Tree Regulations and Their Implementation," *Journal of Arboriculture*, Vol. 4, No. 6 (June 1978): 142.

65. Ralph R. Widner, "Maryland: Water on a Warden's Back," in Ralph R. Widner, ed., *Forests and Forestry in the American States* (Washington, DC: The National Association of State Foresters, 1968), 239.

66. Ibid.

67. "Fred Wilson Besley" (Jane Overington Wright Collection).

68. "Shade Trees Along the Public Highways of State—Beautifying of the State Roads of Maryland Next on Program," *Oakland Republican* (8 April 1920).

69. *Baltimore Sun*, 22 March 1920; 9 and 10 April 1920; 31 March 1921; 2, 21, and 25 April 1921; 21 May 1921; *Baltimore News*, 11, 30, and 31 March 1921; and *Oakland Republican*, 4 March 1920.

70. F. W. Besley, *Report of the State Department of Forestry, University of Maryland, for Fiscal Years 1924 and 1925, October 1, 1923 to September 30, 1925* (Baltimore: State Department of Forestry, 1925)), 49.

71. "F. W. Besley," *Yale Forest School News*, Vol. 11, No. 2 (April 1923): 31.

72. Nelson, "State Recreation," 111.

73. See, also, Berlin *Advertiser* (16 July 1920); Elkton's *Cecil Whig* (22 January 1921); and *Easton Gazette* (18 February 1921).

74. "F. W. Besley," *Yale Forest School News*, Vol. 3, No. 3 (July 1915): 35.

75. "Maryland State Board of Forestry, Forestry Leaflet No. 18, Plan of Co-operation between Woodland Owners and the State Forester," in *Report for 1914 and 1915*, 54–5.

76. F. W. Besley, *Report of the State Department of Forestry, University of Maryland, for 1922 and 1923* (Baltimore: State Department of Forestry, 1923), 7.

77. Forest Laws of Maryland (see note 63).

78. Totals were taken from the Report of the State Board of Forestry for 1912 and 1913 and the Report of the State Board of Forestry for 1920 and 1921.

79. See Chapters 348, 749, and 794 in the *Laws of the State of Maryland: Made and Passed* (Baltimore: King Brothers, State Printers, 1912).

80.   Geoffrey L. Buckley, Robert F. Bailey, and J. Morgan Grove, "The Patapsco For-est Reserve: Establishing a "City Park" for Baltimore, 1907–1941," *Historical Geography*, Vol. 34 (2006): 87–108.

81.   Deeds on file at the Maryland Department of Natural Resources in Annapolis, Maryland.

82.   F. W. Besley, *Yale Forest School News*, Vol. 3, No. 4 (October 1915): 51.

83.   Clark, Physical Features of Maryland, 247–8.

84.   Besley, Maryland's Forest Resources, 8–9.

85.   *Baltimore Sun* (9 February 1921).

86.   Records of the President's Office. Series VII, Box 17. State Forests Acquired as of May 17, 1934. University of Maryland, University Archives, College Park, Maryland.

87.   Besley, *Forests of Allegany County*, 5.

88.   In *State Recreation: Parks, Forests and Game Preserves*, Nelson identifies Maryland as one of twenty-nine states that did not possess a single state park in 1921. If one goes along with Freeman Tilden's definition of a state park, ". . . any area of any size set aside for any type of recreation purpose, or as a historical memorial, or to preserve scenery or a natural curiosity, and *called* a state park," then one could argue that Maryland's emerging system of forest reserves included parcels that met all but the last of these requirements. Freeman Tilden, *The State Parks: Their Meaning in American Life* (New York: Alfred A. Knopf, 1970), 4. See, also, Robert Shankland, *Steve Mather of the National Parks*, 3rd ed. (New York: Alfred A. Knopf, 1951); Ney C. Landrum, ed., *Histories of Southeastern State Park Systems* (Tallahassee: Association of Southeastern State Park Directors, 1992); and Ney C. Landrum, *The State Park Movement in America: A Critical Review* (Columbia: University of Missouri Press, 2004).

89.   J. Gordon Dorrance, *The State Reserves of Maryland: "A Playground for the Public"* (Baltimore: Maryland State Board of Forestry, 1919), 7; Besley, *The Forests of Maryland* (Baltimore: Maryland State Board of Forestry, 1916) and the *Report for 1910 and 1914* (Baltimore: State Board of Forestry, 1911) also make reference to state parks.

90.   Dorrance, ibid.

91.   Dorrance, ibid., 9. Interesting to note, the author states: "This leaflet, as it now appears, is in large part a reprint and extension of one prepared earlier by the Board of Forestry, and published in 1916 by the Baltimore and Ohio Railroad." Certainly, the

B&O hoped to capitalize on the public's recreational use of the reserves by transporting vacationers to both the Patapsco and western forest reserves.

92. *Baltimore Evening Sun* (21 January 1921).

93. Ibid.

94. Interview with Helen Besley Overington.

95. Hal K. Rothman, "'A Regular Ding-Dong Fight': The Dynamics of Park Service-Forest Service Controversy During the 1920s and 1930s," in Char Miller, ed., *American Forests: Nature, Culture, and Politics* (Lawrence: University Press of Kansas, 1997), 109–24.

96. Interview with Kirk Rodgers, 29 November 2003.

97. F. W. Besley, Report of the State Department of Forestry, University of Maryland, for 1922 and 1923.

98. Besley, "Partial Biography," 6.

99. F. W. Besley, *Yale Forest School News*, Vol. 4, No. 4 (October 1916): 51.

Chapter Three

The chapter epigraph comes from Anne Whiston Spirn, *The Granite Garden: Urban Nature and Human Design* (New York: Basic Books, Inc. 1984), 175.

1. Board of Park Commissioners, *Public Parks of Baltimore, No. 2, Patterson Park* (Baltimore: Board of Park Commissioners, 10 December 1927), 5–6; and Lisa Baynes and Douglas B. Brady, "History of Patterson Park" (Baltimore: Department of Recreation and Parks, 1985).

2. J. V. Kelly, *Public Parks of Baltimore, No. 3, Druid Hill Park: The Land and Its People During the Period of Private Ownership* (Baltimore: Board of Park Commissioners, 10 June 1928); J. V. Kelly, *Public Parks of Baltimore, No. 4, Druid Hill Park: Events Resulting in Its Acquirement, and Method of Financing* (Baltimore: Board of Park Commissioners, 10 September 1929); and Bennard B. Perlman, "Druid Hill Park: Centenary," *Baltimore Sun* (19 October 1960).

3. Howard Daniels, "First Annual Report of the Landscape Gardener of Druid Hill Park," in *The First Annual Report of the Public Park Commission to the Mayor and City Council of Baltimore* (December 31st, 1860), 461.

4. Ibid., 454–55.

5.   Ibid., 459.

6.   Ibid., 460.

7.   Howard Daniels, "Second Annual Report of the Landscape Gardener of Druid Hill Park," in *Second Annual Report of the Park Commission of the City of Baltimore, to the Mayor and City Council of Baltimore, For the Year Ending 31st December, 1861* (Baltimore: King Brother & Armiger, Printers), 25–26.

8.   Ibid., 7–8 and 26.

9.   *Eighteenth Annual Report of the Park Commission, to the Mayor and City Council of Baltimore, For the Fiscal Year Ending Dec. 31, 1877* (Baltimore: Printed by John Cox, 1878), 9.

10.   *Nineteenth Annual Report of the Park Commission, to the Mayor and City Council of Baltimore, For the Fiscal Year Ending Dec. 31, 1878,* 6.

11.   *Thirty-First Annual Report of the Public Park Commission to the Mayor and City Council of Baltimore, For the Fiscal Year Ending December 31, 1890,* 21.

12.   *Second Annual Report of the Park Commission* (1861), 9; *Report of the Public Park Commission to the Mayor and City Council of Baltimore* (1865), 496; *Twenty-Sixth Annual Report of the Park Commission to the Mayor and City Council of Baltimore, For the Fiscal Year Ending December 31, 1885,* 37; and *Thirty-First Annual Report of the Public Park Commission* (1890), 21.

13.   *Report of the Public Park Commission* (1865), 496; and *Nineteenth Annual Report of the Park Commission* (1878), 6–7.

14.   *Eighteenth Annual Report of the Park Commission* (1877), 23; *Nineteenth Annual Report of the Park Commission* (1878), 30–32; *Twentieth Annual Report of the Park Commission, to the Mayor and City Council of Baltimore, For the Fiscal Year Ending December 31, 1879,* 16 and 39–40; *Twenty-First Annual Report of the Park Commission to the Mayor and City of Baltimore, For the Fiscal Year Ending December 31st, 1880,* 37–38; and *Twenty-Sixth Annual Report of the Park Commission* (1885), 45.

15.   Olmsted Brothers, *Report Upon the Development of Public Grounds for Greater Baltimore* (Baltimore: Friends of Maryland's Olmsted Parks & Landscapes, Inc., [1904] 1987), 49; and *44th and 45th Annual Reports of the Board of Park Commissioners to the Mayor and City Council of Baltimore, For the Fiscal Years Ending December 31, 1903,* 1904 (Baltimore: Wm. J. C. Dulany Company, City Printers, 1905), 34.

16.   *44th and 45th Annual Reports,* 34–35.

17.   Rev. D. H. Steffens, "What May Be Done Toward Reforesting Baltimore," *Baltimore Morning Sun* (6 October 1912).

18.   Ibid.

19.   Ibid.

20.   *Baltimore Morning Sun* (14 August 1907).

21.   Ibid.

22.   *Baltimore Morning Sun* (17 February 1912).

23.   James H. Preston, Mayor. Ordinance No. 154 of the Mayor and City Council of Baltimore. An Ordinance to Regulate the Planting, Trimming and Removing of Trees, in the Streets of Baltimore; and Providing for a City Forester and Assistants for that Purpose. Approved August 17th, 1912, 1–2.

24.   *Baltimore Morning Sun* (23 August 1912); and Preston, Ordinance No. 154, 1.

25.   *Baltimore Morning Sun* (8 September 1912).

26.   *Baltimore Morning Sun* (13 October 1912).

27.   According to his obituary, "He was a graduate of Baltimore Polytechnic Institute in 1909 and in 1938 received his M.F. degree from Yale as a member of the class of 1913." See *Yale Forest School News*, Vol. 57, No. 2 (October 1969): 34.

28.   Warren Wilmer Brown, *The Municipal Art Society of Baltimore City: Its Aims and Accomplishments* (Baltimore: Municipal Art Society, ca. 1930), 26 and 28.

29.   Sherry H. Olson, *Baltimore: The Building of an American City*, 2nd ed. (Baltimore: The Johns Hopkins University Press, 1997).

30.   Eric L. Holcomb, *The City as Suburb: A History of Northeast Baltimore since 1660*, *Updated Edition* (Chicago: Center for American Places at Columbia College Chicago, 2008), 175–76 and 211. See, also, *The Northeast Baltimore Improvement Association of Baltimore City Constitution* (Baltimore: William T. Hynes, Printer, 1907). A copy is on file at the Maryland Historical Society, Baltimore.

31.   Garrison Boulevard Association, Meeting Minutes, 15 May 1933, 1; President's Reports, 20 October 1930, 2; and 16 October 1933, 1–2. Copies are on file at the Maryland Historical Society, Baltimore.

32.   Constitution of the Peabody Heights Improvement Association of Baltimore City. (Baltimore: Press of A. Hoen & Co., 1908), 4; and Peabody Heights Improvement Association Meeting Minutes, 1909–1933, Book 2, 14 December 1931, 105. Copies on file at the Maryland Historical Society, Baltimore, Maryland

33.   Peabody Heights Improvement Association, Book 1, 14 March 1910, 43; and 9 May 1910, 47–48.

34.   Peabody Heights Improvement Association, Book 1, 10 October 1910, 57–58.

35.   Peabody Heights Improvement Association, Book 1, 13 February 1911, 71–72.

36.   Peabody Heights Improvement Association, Book 1, 12 February 1912, 110.

37.   Peabody Heights Improvement Association, Book 1, 10 November 1913, 173; 14 December 1914, 201; 8 March 1915, 216; 14 June 1915, 225; 11 October 1915, 228–229; 13 December 1915, 231; and 14 June 1920, 307.

38.   Peabody Heights Improvement Association, Book 1, 11 November 1912, 140; 8 December 1913, 177.

39.   Besley's entry for the January 1920, Vol. 8, No. 1 issue of *Yale Forest School News* describes the event: "Experienced serious loss on account of destruction by fire of all of his office records. Fire was at midnight with no one around to save anything."

40.   Peabody Heights Improvement Association, Book 1, 11 November 1912, 138; 12 February 1912, 108; 15 May 1913, 165; and 9 February 1914, 189. For an excellent introduction to the activities of the American Civic Association, see Terence Young, "Social Reform through Parks: The American Civic Association's Program for a Better America," *Journal of Historical Geography* Vol. 22, No. 4 (1996): 460–72.

41.   Peabody Heights Improvement Association, Book 1, 10 October 1910, 58.

42.   Peabody Heights Improvement Association, Book 1, 10 April 1922, 344.

43.   Henry W. Lawrence, "Changing Forms and Persistent Values: Historical Perspectives on the Urban Forest," in Gordon A. Bradley, ed., *Urban Forest Landscapes: Integrating Multidisciplinary Perspectives* (Seattle: University of Washington Press, 1995), 31.

44.   *The Mount Royal District: Baltimore's Best Urban Section: Dolphin Street to Druid Hill Park between Mount Royal Ave. and Eutaw Place.* Protected by the Mount Royal Improvement Association. An incorporated body of property owners, maintaining a permanent office with a full time secretary (Baltimore: Mount Royal Improvement Association, 1930), 5. A copy is on file at the Maryland Historical Society, Baltimore.

45.   *The Mount Royal District*, 6, 12. See, also, Cynthia L. Merse, "Historical Geography of Urban Forestry and Roadside Tree Planting in Baltimore" (M.S. thesis, Ohio University, 2005).

46.   *The Mount Royal District*, 39.

47.   NOTICE OF MEETING, A special meeting of the MOUNT ROYAL IMPROVEMENT ASSOCIATION will be held at the ASSOCIATE CONGRE-GATIONAL CHURCH Northwest corner of Maryland Avenue and Preston Street Tuesday, June 24th at 8 P.M. (Daylight Saving Time) [1930?]. A copy is on file at the Maryland Historical Society, Baltimore, Maryland.

48.   Olson, *Baltimore: The Building of an American City*, 273.

49.   "R.B. Maxwell, 79, Was Parks Chief," *Baltimore Evening Sun* (4 March 1969); *Yale Forest School News*, Vol. 5, No. 3 (July 1917): 46.

50.   R. Brooke Maxwell, First Annual Report of the City Forester of the City of Baltimore to the City Engineer for the Fiscal Year Ending December 31, 1913 (Baltimore: Meyer & Thalheimer, 1914).

51.   Ibid., 6–8.

52.   Ibid., 8–11.

53.   Ibid., 12 and 14–15.

54.   Ibid., 9.

55.   Ibid., 15.

56.   *Reports of the City Officers and Departments Made to the City Council of Baltimore for the Year 1914* (Baltimore: Meyer & Thalheimer, 1915), 16.

57.   Report of R.B. Maxwell, City Forester, Sub-Division of Forestry *Reports of the City Officers and Departments Made to the City Council of Baltimore for the Year* (Baltimore: Meyer & Thalheimer, 1916), 89.

58.   Ibid., 90.

59.   *Reports of the City Officers and Departments Made to the City Council of Baltimore for the Year 1916* (Baltimore: Meyer & Thalheimer, 1918), 79.

60.   Ibid., 79.

61.   Ibid., 80.

62.   Ibid., 81.

63.   Ibid., 82–85.

64.   R. Brooke Maxwell, "News from the Baltimore Arborist" *Yale Forest School News*, Vol. 5, No. 1 (January 1917): 9.

65.   *Yale Forest School News*, Vol. XXXV, No. (January 1947): 12.

66.   Report of Carl Schober, City Forester, 1917, Highways Engineer's Department, Sub-Division of Forestry. Baltimore, April 8th, 1918. *Reports of the City Officers and Departments Made to the City Council of Baltimore for the Year 1917* (Baltimore: King Bros. 1918), 75.

67.   Ibid., 75–77.

68.   Ibid., 78–79.

69.   *Annual Report, Department of Legislative Reference. Forestry Division. To the Mayor and City Council of Baltimore For the Fiscal Year Ending December 31, 1918* (Baltimore: King Bros. 1918), 126.

70.   *Reports of the City Officers and Departments Made to the City Council of Baltimore for the Year 1919* (Baltimore: King Bros. 1920), 138.

71.   Maxwell, *First Annual Report of the City Forester*, 10–11.

72.   *Report of the Park Commission to the Mayor and City Council of Baltimore* (1920), 44.

73.   Peabody Heights Improvement Association, Book 2, 9 January 1928, 58.

74.   *Report of the Park Commission* (1920), 44.

75.   "Beauties of Halethorpe Compared with Guilford's: Back and Front Yards Turned Into Miniature Orchards—Arborial Growths Get Expert Care," *Baltimore Evening Sun* (28 May 1921). It should be noted here that City Forester Maxwell also engaged in the practice of planting nut-bearing trees, as this passage from a *Baltimore Morning Sun* article (dated March 6, 1916) shows: "With the approval of the Water Board, City Forester R. Brooke Maxwell will inaugurate early in the spring a plan under which nut-bearing trees will be planted along the sidewalks, probably in front of city property. He has arranged to place 12 trees, either English walnut or pecan, on the Malster street side of the Mount Royal pumping station. 'These plantings,' said Mr. Maxwell yesterday, 'will mark the beginning of a plan to plant trees for something more than shade, following the custom in other cities. . . .' The nut-bearing trees were in a list of 620 trees of various kinds Mr. Maxwell's department will plant in different parts of the city between March 15 and May 10, with the sanction of the Board of Estimates, given last week, it was learned yesterday."

76.   Ibid.

77.   *Baltimore Evening Sun* (22 August 1946).

78.   *City of Baltimore, Maryland, Department of Public Parks and Squares, Annual Report 1946 (To the Board of Park Commissioners for the Year Ending December 31, 1946)*, 7.

Chapter Four

The chapter epigraphs come from W. B. Greeley, "Future Trends in National and State Forestry," *American Forests*, Vol. 33, No. 397 (January 1927): 3; and F. W. Besley, "State Forests and Parks," in W. S. Hamill, *The Forest Resources and Industries of Maryland* (Baltimore: Maryland Development Bureau of the Baltimore Association of Commerce, 1937), 184–86.

1.   *Report of the State Department of Forestry, University of Maryland, for 1922 and 1923* (Baltimore, Maryland), 7–8.

2.   *Report of the State Department of Forestry, University of Maryland, for Fiscal Year 1926, October 1, 1925 to September 30, 1926* (Baltimore, Maryland), 15.

3.   *Report of the State Department of Forestry, University of Maryland, for Fiscal Year 1927, October 1, 1926 to September 30, 1927* (Baltimore, Maryland), 7–8.

4.   *Report for 1922 and 1923*, 9.

5.   *Report for 1927*, 8.

6.   James Chace, 1912: *Wilson, Roosevelt, Taft & Debs—The Election That Changed the Country* (New York: Simon & Schuster, 2004), 279.

7.   Neil M. Maher, "'A Conflux of Desire and Need': Trees, Boy Scouts, and the Roots of Franklin Roosevelt's Civilian Conservation Corps," in Henry L. Henderson and David B. Woolner, eds., *FDR and the Environment* (New York: Palgrave MacMillan, 2005), 52.

8.   *Report for 1926*, 7.

9.   Edna Warren, "Forests and Parks in the Old Line State," *American Forests*, Vol. 62, No. 10 (October 1956): 22.

10.   Ibid., 23.

11.   Ibid.

12.   Ibid., 25.

13.   Report for 1926, 7.

14.   Maryland State Department of Forestry, *Report by the Advisory Board of Forestry,* (Baltimore, Maryland), 8.

15.   Ibid., 9.

16.   *Report for 1926*, 7. The responsibilities of forest wardens were clearly spelled out in the *Maryland Tree Wardens Manual*, published by the State Department of Forestry in

1929, a copy of which can be found at the Enoch Pratt Free Library in Baltimore.

17. *Report of the State Department of Forestry, University of Maryland, for Fiscal Year 1930, October 1, 1929, to September 30, 1930* (Baltimore, Maryland), 26.

18. *Report of the State Department of Forestry, University of Maryland, for Fiscal Years 1924 and 1925, October 1, 1923 to September 30, 1925* (Baltimore, Maryland), 11.

19. *Report of the State Department of Forestry, University of Maryland, for Fiscal Years 1931 and 1932, October 1, 1930, to September 30, 1932* (Baltimore, Maryland), 7.

20. *Report for 1930*, 42.

21. Ibid., 41–44.

22. Ibid., 41–42.

23. Ibid., 42.

24. Ibid., 27.

25. Ibid., 43.

26. Ibid., 27.

27. *Report for 1926*, 32.

28. *Report for 1930*, 26.

29. *Report for 1940*, 9.

30. *Report of the Commission of State Forests and Parks, 1943–1944* (Annapolis, MD, 1945), 9.

31. Ibid., 7–8.

32. Warren, "Forests and Parks," 22.

33. *Report for 1922 and 1923*, 18. In the State Department of Forestry *Report for 1929* (p. 10), Besley uses U.S. census figures to drive his point home: "[F]rom 1900 to 1925 there was a decrease of 538,110 acres in the improved farm acreage in Maryland, amounting to 16.6 per cent. Between 1920 and 1925 the decrease was 143,486 acres, or about 4½ per cent. The acreage in farm crops during the 5-year period, 1920-1925, decreased 454,545 acres, or a drop of over 20 per cent., while during the same period the decrease in farm acreage was less than 7 per cent. These figures show that for many years the cultivated acreage for crop production has been steadily decreasing, with a corresponding increase in non-tilled land. It is probable that some of the land withdrawn from farm acreage has gone into suburban development, parks and the like, but the decrease in existing farm acreage that is no longer used for field crops represents mainly idle land, which is reforesting naturally even though imperfectly, and which should be planted in forest."

34. *Report for 1924 and 1925*, 8.

35. According to the *Report for 1930* (p. 21): "[T]he State Forester and members of his staff gave during the year 121 lectures, 78 of which were illustrated. These were under the auspices of schools, colleges, Parent-Teacher Associations and other civic associations."

36. *Report for 1931 and 1932*, 9.

37. *Report for 1922 and 1923*, 8, 17.

38. *Report for 1924 and 1925*, 12.

39. *Report for 1928*, 7.

40. *Report for 1929*, 8.

41. *Report for 1924 and 1925*, 14.

42. *Report for 1926*, 8.

43. *Report for 1927*, 15.

44. *Report for 1924 and 1925*, 15.

45. Ibid. See also *Report for 1928*, 12.

46. *Report for 1927*, 15.

47. *Report for 1927*, 16; and *Report for 1929*, 19.

48. *Report for 1926*, 18.

49. *Report for 1930*, 15.

50. *Report for 1924 and 1925*, 18; and *Report for 1926*, 19.

51. *Report for 1930*, 11–12.

52. *Report for 1931 and 1932*, 9.

53. *Report for 1932*, 9.

54. Records of the President's Office, Series VII, Box 17. Recommendation from the Florists Club of Baltimore enclosed with a letter from William C. Price of Towson Nurseries to Dr. R. A. Pearson (February 9, 1932), University of Maryland, University Archives, College Park, Maryland.

55. Records of the President's Office. Series VII, Box 17, Memorandum: A Forest Nursery (March 24, 1932), University of Maryland, University Archives, College Park, Maryland.

56. Ibid.

57. Records of the President's Office, Series VII, Box 17, Proposed Rules for Distribution of Planting Stock from State Forest Nursery Effective July 1, 1932, University of Maryland, University Archives, College Park, Maryland.

58. Records of the President's Office. Series VII, Box 17, Letter from F.W. Besley to H. Street Baldwin (December 13, 1934), University of Maryland, University Archives, College Park, Maryland.

59. Ibid.

60. Records of the President's Office, Series VII, Box 17, Letter from F. W. Besley to Dr. Pearson, President, University of Maryland (March 3, 1934), University of Maryland, University Archives, College Park, Maryland; and Records of the President's Office, Series VII, Box 17, Letter from William Price to Dr. Pearson, President, University of Maryland (March 10, 1934), University of Maryland, University Archives, College Park, Maryland.

61. *Report for 1924 and 1925*, 18; Francis Zumbrun, Kathy Kronner, and Ethan Kearns, "Big Tree Champions (www.dnr.state.md.us/centennial/BigTreeChampions_History.asp); and author's interview with Kirk Rodgers, 30 November 2007. See, also, Barbara Bosworth, *Trees: National Champions* (Cambridge: The MIT Press, 2005).

62. *Report for 1932*, 7. The decline in assistance to woodland owners is clearly reflected in the figures Besley kept. During 1924 and 1925, 107 examinations were carried out in twenty of Maryland's twenty-three counties covering some 9,052 acres of woodland (*Report for 1924 and 1925*, 8). In 1926, fifty-one such examinations were performed on 11,500 acres (*Report for 1926*, 7). Two years later, 100 examinations covering 8,761 acres were recorded (*Report for 1928*, 8). By 1930, the figures had dropped to forty-two examinations comprising 3,199 acres (*Report for 1930*, 8). The numbers dropped even further in 1931—twenty-six tracts totaling 1,481 acres (*Report for 1932*, 8).

63. *Report for 1922 and 1923*, 8.

64. *Report for 1926*, 12.

65. *Report for 1924 and 1925*, 10. Besley suggested that the state should control "at least 150,000 acres, and the municipalities 50,000."

66. *Report for 1926*, 12.

67. Ibid., 14.

68. Ibid., 12.

69. Ibid., 13.

70. *Report for 1928*, 8.

71. *Report for 1929*, 8.

72. *Report for 1927*, 22.

73. Ibid. There can be little doubt that Besley believed in local control. As the following passage indicates (*Report for 1927*, 18–19), he was even willing to cede control

in some cases: "The City of Baltimore for many years has had a City Forester who is given exclusive jurisdiction over shade-tree matters within the city limits. The City of Frederick has recently appointed C. Cyril Klein, formerly District Forester in the State service, as City Forester of Frederick. Since his appointment all public shade-tree work in the City of Frederick has been handled by the City Forester to the satisfaction of both the city administration and the Department of Forestry. The growing interest in parks and shade trees within the city limits demands more and more attention and the appointment of a City Forester seems to be the proper solution of this administrative problem. Local self-government in these matters is highly desirable and, where a man with proper training can be secured and provided with a working budget, it is the most satisfactory procedure."

74.   *Baltimore Sun* (9 December 1920).

75.   W. B. Greeley, "Future Trends in National and State Forestry," *American Forests and Forest Life*, Vol. 33, No. 397 (January 1927): 2.

76.   Ibid., 6.

77.   Ibid., 8.

78.   Maryland's Strange Reasoning," *American Forests and Forest Life*, Vol. 33, No. 403 (July 1927): 421.

79.   Ibid., 422.

80.   Ibid.

81.   It was not a new argument. On December 17, 1925, Besley touched on many of the same topics in a paper delivered at the annual meeting of the Society of American Foresters in Madison, Wisconsin, entitled "State Forests in Relation to the National Forest Program." Among other things, he noted that state forests "help to stabilize and maintain wood-using industries"; allow for the demonstration of "correct principles of forest management"; facilitate reduction of forest fires; protect watershed and "water-power development"; and "furnish a shelter and breeding place for game." He also identified other "supplemental values" that state forests provide; namely, recreation. Using the Adirondacks as an example, he pointed out that state forests can serve as "pleasure forests and recreational grounds for the enjoyment of the people." Perhaps he was alluding to his own situation in Maryland when he likened state forestry departments without state forest lands to manage to "a theoretical farmer, owning no lands of his own, but attempting to tell his neighbors how to run their farms."

82. F. W. Besley, "State Forests vs. National Forests: Maryland's Answer," *American Forests and Forest Life*, Vol. 33, No. 404 (August 1927): 470.

83. Ibid.

84. Ibid. According to George H. Callcott, "People in Maryland, like people in most states, never came to terms politically with the depression and the New Deal, and they accepted the controls and services the war brought only for purposes of winning the war. From depression to war and on into the postwar world, people were caught between the philosophical rejection of big government and the need for more of the things it provided. The shadow over Maryland politics was Albert C. Ritchie, the four-term Democratic governor (1920-1935), a symbol of best-government-is-least conservatism. He considered Coolidge and Hoover to be spendthrifts and he periodically surfaced as a possible Democratic presidential candidate who might attack the Republicans from the right. . . . Maryland voted by landslides for Roosevelt and accepted the handouts Washington provided, but through all of Roosevelt's administrations, Maryland's governors, its United States senators, its Baltimore mayors, and the majority of its legislators offered lip-service support at best, and in practice they were usually uncooperative," from *Maryland & America, 1940 to 1980* (Baltimore: The Johns Hopkins University Press, 1985): 52.

85. Ibid.

86. F. W. Besley, "State and Federal Acquisition of Forest Lands in the East," *Journal of Forestry*, Vol. XXVII, No. 2 (February 1929): 113.

87. Ibid., 113–14.

88. Ibid., 114–15.

89. Ibid., 117. Besley's entry for the July 1935 issue of *Yale Forest School News* (p. 44) suggests a compromise. Federal funds could be used to purchase state forests if the states agreed to administer them and repay the federal government at a later date: "Fred Besley acted as spokesman for the Association of State Foresters at a hearing on House Bill 6914, held on April 11th, before a subcommittee of the House Agricultural Committee. This hearing was conducted in order to organize backing for the proposed program calling for a federal appropriation of $20,000,000 for the purchase of lands suitable for state forests. It is proposed that these lands be purchased by the government and turned over to the states for administration, their title to remain in the government until the states can repay the original purchase price from 50 per cent of the receipts obtained through operation of the forests."

90.   Records of the President's Office, Series VII, Box 17, Letter from F.W. Besley to Dr. R.A. Pearson, President, University of Maryland (December 17, 1929), University of Maryland, University Archives, College Park, Maryland.

91.   *Report for 1930*, 13; and Letter from Besley to Pearson, December 17, 1929.

92.   Records of the President's Office, Series VII, Box 17, Letter from F.W. Besley to the Board of Regents, University of Maryland (July 13, 1931), University of Maryland, University Archives, College Park, Maryland. With respect to "Demonstration Forests," the case involving the Seth property on the Eastern Shore is illustrative. In 1928, Mary W. Seth, of Easton, "deeded a tract of 65 acres to the Department as a memorial to her husband, General Joseph B. Seth, who died in November, 1927. The stipulation was that it should be known as the Seth Demonstration Forest and should be used to demonstrate proper forestry practices" (*Report for 1928*, 11).

93.   Records of the President's Office, Series VII, Box 17.

94.   Ibid.

95.   *Report for 1930*, 13–14.

96.   Fred W. Besley, "State Forests and Parks," in *The Forest Resources and Industries of Maryland* (Baltimore: Maryland Development Bureau of the Baltimore Association of Commerce, 1937), 184–85.

97.   Ibid., 185.

98.   *Report for 1930*, 13.

99.   *Report for 1931 and 1932*, 11.

100. Records of the President's Office, Series VII, Box 17, Unpublished memorandum to President Pearson dated 16 May 1932 and titled "Maryland's Forest Resources" (May 16, 1932), University of Maryland, University Archives, College Park, Maryland. Besley's entry for the April 1932 issue of *Yale Forest School News* confirms these goals: "In spite of the depression, our State Forest Acquisition program is moving ahead. We now have 15,000 acres under option, with intent to purchase. With the 35,000 acres already acquired, this should give us about 50,000 acres by the end of this year. We hope to have 200,000 acres eventually, nearly 10 per cent of the forest area of the State. This is perhaps rather high for an agricultural state like Maryland, consisting largely of woodlots."

101. Besley, "State Forests and Parks," 186.

102. *Report for 1931 and 1932*, 11.

103. *Report for 1940*, 4.

104. Ibid., 17.

105. Letter from F. W. Besley to H. C. Buckingham, 17 December 1935. A copy is on file at the Green Ridge State Forest Headquarters.

106. Letter from H.C. Buckingham to Resident Forest Wardens, 10 August 1932. A copy is on file at the Green Ridge State Forest Headquarters.

107. *Report for 1940*, 6.

108. According to the *Report for 1940* (p. 4): "The last five years have brought about a marked change in the relationship of the public to the State Forests. Construction of roads and trails by the Civilian Conservation Corps has opened these lands to public use, and the people have come in ever-increasing numbers. In 1939, aside from the State parks, more than 100,000 persons visited the State Forests. One potent source of attraction was the New Germany recreational area, on the Savage River State Forest, in Garrett County. Here, under the direction of the State Forester, the CCC has developed a 10-acre artificial lake, and nearby has built recreation structures and rustic log cabins, the latter for public rental, together with the necessary water and sewage disposal systems. A commodious stone bath house, a sand beach, and additional cabins are under construction. Unique at New Germany are the ski trails, both for beginners and for experts, that have attracted wide attention and have been enjoyed by people from every part of Maryland and from practically every adjacent State. On a single Sunday last winter, 460 cars were counted at New Germany. In the summer, of course, visitors are even more numerous. For July and August, 1939, the figures were 8356 and 5436, respectively. A similar, and even more extensive, recreational development is under way at Herrington Manor, on the Swallow Falls State Forest, where a 55-acre lake has been constructed. On this same Forest, and giving it its name, is Swallow Falls, one of the most beautiful spots in the State, with is surrounding stands of virgin white pine and hemlock, last of the mighty timber that once clothed the mountains of western Maryland. Adequate picnic grounds and other recreational facilities have been developed here, and on pleasant week-ends as many as 2500 people visit the spot." The CCC contributed to the recreational development of other areas in the state as well. For a more detailed examination of the CCC's role in Maryland, see Robert F. Bailey, *Maryland's Forests and Parks: A Century of Progress* (Charleston, SC: Arcadia Publishing, 2006).

109. *Report for 1940*, 17.

110. In making the case for purchasing a tract of land on the Eastern Shore owned by Coulbourn Bros. of Cape Charles, Virginia, Besley described the condition of the parcel

he was interested in acquiring: "This land is in Worcester County and while it does not border the present State Forest, it is less than one mile distant, separated by another tract of considerable size which can likely be purchased at a reasonable cost. It was cut-over about sixteen years ago and is now stocked with a good young growth, mainly loblolly pine, and rated as excellent timber growing land, easily accessible and very desirable for State Forest purposes. A careful examination has been made of the land and it is highly recommended as a very advantageous purchase." Records of the President's Office, Series VII, Box 17, Letter from F. W. Besley to Mr. George W. Shriver, Chairman, Board of Regents, University of Maryland (February 27, 1934), University of Maryland, University Archives, College Park, Maryland.

111. Records of the President's Office, Series VII, Box 17, Letter from F. W. Besley to Dr. R. A. Pearson, President, University of Maryland (January 23, 1935), University of Maryland, University Archives, College Park, Maryland. In 1925, Besley observed (in "State Forests in Relation to the National Forest Program"): "In most of the states, state parks and state forests are very closely coordinated, in many cases under the same management. This seems to be a logical method of procedure and one calculated to cause the least friction or duplication of effort. The state forestry departments realize that areas on state forests that are better adapted for park use can be set aside for this purpose and administered as such. The uses of these areas are determined largely by local needs and those who administer the state forests are generally responsive to such needs. It usually means simply the setting aside of areas of high scenic or recreational value for park use, and the development of other areas for timber production, and such areas can be administered much more economically under the forest administration than to have them set apart and administered separately under a park commission. Besley followed this up in 1937 (in "State Forests and Parks," 185): "It is important that the acquisition of State forests and parks constitute a joint program, without duplication of effort and competitive bidding for lands."

112. Letter from F. W. Besley to George M. Shriver, Chairman, Board of Regents (March 20, 1935).

113. Letter from F. W. Besley to Mr. Lawrence C. Merriam, Acting Assistant Director, National Park Service (July 5, 1935), National Archives II, College Park, Maryland.

114. Warren, "Forests and Parks," 56. Reminiscing in 2004, Besley's daughter, Helen Overington, recalled that her father was not particularly happy with the selection of Kaylor as his replacement. Apart from the fact that he favored the candidacy of his

long-time friend and colleague, Karl Pfeiffer, he did not approve of Kaylor's emphasis on parks. For a less sympathetic interpretation of Besley's position on parks vis-à-vis forests, see Eugene P. Parker, "When Forests Trumped Parks: The Maryland Experience, 1906-1950," *Maryland Historical Magazine*, Vol. 101, No. 2 (Summer 2006): 203–24.

115. *Yale Forest School News*, Vol. 13, No. 2 (April 17, 1925): 25.

116. In 1927, for example, Besley wrote at least three letters requesting an increase in pay for forest wardens. This one is dated August 15, 1927: "As a justification for granting an increase in pay to the forest wardens engaged in tree work: These men are invested with great authority and are responsible for the conduct of important work, where knowledge of the subject, good judgment and dependency are required. These men represent the public interest in the roadside and street trees, and stand between the property owners and the pole line companies. When this law was enacted in 1914, there was a great amount of hostility between property owners and the pole line companies over the shade tree problem. This was the situation confronting the Forestry Department in putting the law into effect, and naturally it took sometime to overcome the prejudices of property owners as to the control of the pole line companies. Very soon conditions began to improve. The property owners realized that the tree warden stood for their own best interests as against the pole line companies who, naturally, would like to see all trees along the highways interfering with their lines removed. The tree wardens receive their instructions from and are paid by the Forestry Department, and report directly to it, so that they can act in an independent way. . . . Considering the diversity of interest involved, the work has moved along with surprising smoothness, giving general satisfaction. We are endeavoring to maintain a high standard of service, and in order to do so, we must select men of character and ability, and they should be properly compensated." Records of the President's Office. Series VII, Box 17, Letter from F. W. Besley to Dr. R. A. Pearson, President, University of Maryland (August 15, 1927), University of Maryland, University Archives, College Park, Maryland.

117. F. W. Besley, ca. 1956, "Partial Biography of Fred Wilson Besley" (Kirk Rodgers Collection), 6.

118. Ibid.

119. Records of the President's Office. Series VII, Box 17, Letter from F.W. Besley to Major E. Brooke Lee, Chairman of the Budget Committee of the Board of Regents (December 7, 1932), University of Maryland, University Archives, College Park, Maryland.

120. Besley, "Partial Biography."

121. According to the *Yale Forest School News* for July 1933, Vol. 21, No. 3 (p. 49), Besley was in a unique position to make such an evaluation, having been "Representative of the Director of Emergency Conservation Work." As such he directed" the movement of men from the military conditioning camps to the forest camps in the Third Corps Area of the War Department, which consists of Virginia, Maryland, District of Columbia, and Pennsylvania."

122. Letter from Besley to Pearson, January 23, 1935.

123. Ibid.

124. Records of the President's Office. Series VII, Box 17, Letter from F.W. Besley to Honorable Albert C. Ritchie (September 30, 1933), University of Maryland, University Archives, College Park, Maryland.

125. Ibid.

126. Letter from Besley to Shriver, February 27, 1934.

127. In the *Report for 1940* (p. 11), Besley identified the four sources that supported the work of the Department of Forestry: a state appropriation known as the "General Fund;" a forest reserve fund, which included "all income received by the Department from sources other than the State treasury, such as fines and receipts from the sale of products from State Forests and Parks"; income "from permits issued under the roadside tree law," and Clarke-McNary Act funds received from the federal government. Out of a total budget of $120,594, the state contribution amounted to about $60,594.

128. *Report for 1940*, 14.

129. Besley was also upset that Congress had not completely lived up to its commitment under the terms of the Clarke-McNary Act. In the *Report for 1940* (pp. 11–12) he explained: "The Clarke-McNary Act provides for an annual allotment of Government funds to individual States for the development of fire protection systems, and for expenses incurred in raising trees for forest planting, to be sold to landowners at cost. In the Act it is stipulated that Federal appropriations shall match dollar for dollar the amounts expended by the States for the purposes enumerated. Congress, however, has never appropriated annually more than a fraction of the money to meet State expenditures. For fire protection the Department spends annually [from State funds' the sum of $42,965.00. Of Clarke-McNary funds it receives annually between $10,000.00 and $14,000.00. . . . Unless the Board is mistaken, this is contrary to the provisions of Section 52, Article III, of the State Constitution."

130. *Report for 1940*, 13.

131. Besley, "Partial Biography," 7.

132. Ibid.

133. "Those Among Us You Should Know," *American Forests*, Vol. 37, No. 3 (March 1931): 179.

134. Ibid.

135. Ibid.

Chapter 5

The chapter epigraphs are from remarks by Governor McKeldin, 50th Anniversary Dinner of the State Department of Forestry (Now Forests and Parks), Baltimore, February 16, 1956; and "City Boasts Quarter Million Trees; Fells More in Year Than Is Able to Plant," *Baltimore Evening Sun* (24 May 1955).

1.   Ernest R. Furgurson, "City Losing More Trees Than It's Replacing," *Baltimore Sun* (7 July 1956).

2.   Program for Testimonial Dinner to Honorable Fred W. Besley in Recognition of Thirty-six Years Service as State Forester, In Conjunction with the Annual Dinner Dance of the Maryland State Game and Fish Protective Association, Saturday Evening, February 21, 1942.

3.   Like Greeley, Chapman was also a distinguished member of the Yale School of Forestry's Class of 1904.

4.   Program for Testimonial Dinner.

5.   Ibid. Exhibiting a penchant for understatement, Besley's entry for the July 1913 issue of *Yale Forest School News* (Vol. 1, No. 3) simply states: "Received Honorary Degree of Doctor in Science from Maryland Agricultural College" (p. 30).

6.   *Yale Forest School News*, Vol. 32, No. 3 (July 1944): 52. Fred Besley arrived at West Virginia University on 1 January 1943 as an associate professor of Forest Management. According to the editors of *The Cruiser* (Morgantown: West Virginia University, 1943, p. 17), Besley's advanced age did not slow him down in the least: "It is no wonder that Lowell is agile when his father covers these West Virginia Hills in leaps and bounds." Lowell Besley was also listed as an associate professor of Forest Management in this issue of *The Cruiser*. He had arrived in Morgantown in 1937 after earning a B.S.F. from Cornell

University in 1931 and a M.F. from the Yale School of Forestry in 1932 and after working for the U.S. Forest Service, the Duke Forest, and at Penn State (p. 16). The October 1931 issue (p. 71) of the *Yale Forest School News* (Vol. 19, No. 4) singles out Lowell's entry into the program in the fall of 1931 as particularly noteworthy as he was "the first of the second generation of Yale foresters to do so."

7.   F. W. Besley, "Progress in Forestry," *The Cruiser* (Morgantown: West Virginia University, 1943), 7–8.

8.   Ibid., 8.

9.   Ibid.

10.   *Yale Forest School News*, Vol. 31, No. 2 (April 1943): 27.

11.   *Yale Forest School News*, Vol. 34, No. 2 (April 1946): 28–29.

12.   James Mallow, "Maryland's State Foresters, 1955 to 2001," a paper presented at the annual meeting of the Maryland Forests Association at Rocky Gap State Park, 5 November 2005.

13.   Author's interview with Ross Kimmel, 17 July 2007.

14.   Mallow, "Maryland's State Foresters."

15.   Author's interview with Offutt Johnson, 6 July 2007.

16.   Mallow, "Maryland's State Foresters."

17.   Kimmel, 17 July 2007.

18.   Author's interview with James Mallow, 21 September 2007.

19.   Johnson, 6 July 2007.

20.   Robert F. Bailey, III on behalf of the Maryland Department of Natural Resources. *Maryland's Forests and Parks: A Century of Progress* (Charleston, SC: Arcadia Publishing, 2006), 73.

21.   Mallow, "Maryland's State Foresters."

22.   Ibid.

23.   Bailey, *Maryland's Forests and Parks*, 73.

24.   Author's interview with Kirk Rodgers, 29 November 2003.

25.   Kimmel, 17 July 2007.

26.   Mallow, "Maryland's State Foresters."

27.   Bailey, *Maryland's Forests and Parks*, 93. One of the key sources of funding from the government came from the Land and Water Conservation Fund Act of 1965, which

was passed "in order to preserve, develop, and assure accessibility to outdoor recreation resources." As quoted in Carolyn Merchant, *The Columbia Guide to American Environmental History* (New York: Columbia University Press, 2002), 261.

28. Rodgers, 29 November 2003. According to Rodgers, some land transactions negotiated by Procter and Fred illustrate how they worked together. The Kennedy tract purchase serves as an example. Apparently, Kennedy did not want to sell his land because he knew he had a valuable stand of timber on it. When their offer was declined, Procter suggested they take a different approach. They told Kennedy to cut the timber and then sell the land to them. The strategy was reminiscent of the one Besley employed in 1935, when he requested funds to purchase the Coulbourn tract. In the end, Kennedy sold the land with the valuable timber. As Rodgers likes to say, this piece of land put him through college: "This was their first big sale and it brought in over $100,000."

29. *Yale Forest School News*, Vol. 35, No. 2 (April 1947): 30.

30. "Rites Held for Besley, First State Forester," *Baltimore Evening Sun* (10 November 1960).

31. *Yale Forest School News*, Vol. 35, No. 2 (April 1947): 30.

32. *Yale Forest School News*, Vol. 40, No. 3 (July 1952): 47.

33. *Yale Forest School News*, Vol. 45, No. 4 (October 1957): 64.

34. Author's interview with Kirk Rodgers, 17 December 2004. According to Rodgers, the cabin was built at this location so the family would have a "foothold" on the Eastern Shore. The cinder-block foundation was laid by May, and the house was "under roof" by October. In between, Fred, Procter, and Kirk survived the great hurricane of 1949 and numerous "disagreements" concerning design and construction. The cabin was oriented to take advantage of the southeast winds. The garage, added later, was nicknamed the "gun house" because its construction was financed via the sale of an original Samuel Colt gun. It was sold to family friend and collector Henry Milo. The furniture—mess table, bunks, and chairs—were all U.S. Navy surplus.

35. "R. B. Maxwell, 79, Was Parks Chief," *Baltimore Evening Sun* (4 March 1969).

36. "First State Forester Fred W. Besley Honored at Golden Anniversary Celebration," *The Old Line Acorn*, Vol. 13, No. 1 (March 1956): 4; and *Yale Forest School News*, Vol. 44, No. 2 (April 1956); Vol. 44, No. 3 (July 1956): 56.

37. From Remarks by Governor McKeldin, 50th Anniversary Dinner of State Department of Forestry (Now Forests and Parks), Baltimore, February 16, 1956.

38. Johnson, 6 July 2007.

39. Kimmel, 17 July 2007.

40. Johnson, 6 July 2007.

41. Author's interview with Francis Zumbrun, 23 August 2007.

42. Rodgers, 17 December 2004.

43. *City of Baltimore, Maryland, Department of Public Parks and Squares, Annual Report 1946 (To the Board of Park Commissioners for the Year Ending December 31, 1946)*, 6.

44. *Ibid.*, 7.

45. *Baltimore Evening Sun* (22 August 1946).

46. The January 1947 issue of the *Yale Forest School News* (Vol. 35, No. 1, p. 12) commented on the challenges that awaited Maxwell in his new position: "Judging from newspaper accounts he inherited some difficult personnel and political problems."

47. *City of Baltimore, Maryland, Department of Recreation and Parks, Annual Report* (1951), 18.

48. "Beautify City," *Baltimore News-Post* (22 March 1955).

49. *City of Baltimore, Maryland, Department of Recreation and Parks, Annual Report* (1953), 13.

50. Martin Millspaugh, "City Boasts Quarter Million Trees; Fells More In Year Than Is Able To Plant," *Baltimore Evening Sun* (24 May 1955).

51. Ibid.

52. Furgurson, "City Losing More Trees Than It's Replacing"; "R. Brooke Maxwell," *Yale Forest School News*, Vol. 34, No. 2 (April 1946): 31.

53. "R. Brooke Maxwell."

54. Furgurson, "City Losing More Trees Than It's Replacing," 31.

55. Robert G. Breen, "Containment Of The Elm," *Baltimore Sun* (19 September 1960); "Tree Group Reports On City's Sycamores," *Baltimore News-Post* (16 June 1958).

56. Tom Stevenson, "City Losing Shade Trees: Drastic Program Urged to Save Greenery," *Baltimore News-Post* (10 April 1958).

57. Ibid.

58. "Let's Replace Our Lost Trees," *Baltimore News-Post* (11 April 1958).

59.   Tom Stevenson, "Tree Inquiry Proposed," *Baltimore News-Post* (11 April 1958). Brooke Maxwell may have been the inspiration behind the Women's Civic league campaign. According to the *Baltimore Evening Sun* for March 4, 1969, Maxwell had "planned to plant 1,000 trees a year for 10 years to 'make Baltimore more attractive.' Hearing of the idea, the Women's Civic League adopted it as the theme of the 1955 Flower Mart."

60.   "Mayor Names Group To Study City Trees," *Baltimore News-Post* (undated 1958).

61.   "5-Point Plan Advised In City Tree Program: Recommendations Follow Two Weeks of Study," *Baltimore News-Post* (9 June 1958).

62.   "Tree Committee Action Will Benefit City," *Baltimore News-Post* (9 June 1958).

63.   "'We Want Trees,' Plea to Mayor," *Baltimore News-Post* (25 June 1958).

64.   *Department of Recreation and Parks, City of Baltimore, Maryland, Annual Report* (1957), 33.

65.   *Annual Report of the Department of Recreation and Parks, City of Baltimore, Maryland* (1958), 35–36.

66.   *Annual Report of the Department of Recreation and Parks, City of Baltimore, Maryland* (1959), 34–36.

67.   *Annual Report of the Department of Recreation and Parks, City of Baltimore, Maryland* (1960), 30.

68.   *Annual Report of the Department of Recreation and Parks, City of Baltimore, Maryland* (1961), 36.

69.   *Annual Report of the Department of Recreation and Parks, City of Baltimore, Maryland* (1962), 35–36; *Annual Report of the Department of Recreation and Parks, City of Baltimore, Maryland* (1963), 32; and *Annual Report of the Department of Recreation and Parks, City of Baltimore, Maryland* (1964), 35.

70.   Frank P. L. Somerville, "City's Tree Plans Given: 8,000 To Be Planted Along Streets This Year," *Baltimore Sun* (19 February 1965).

71.   *Annual Report of the Department of Recreation and Parks, City of Baltimore, Maryland* (1965), 30–31; Simonds and Simonds, Landscape Architects and Planners, "A Parks and Recreation Plan for the City of Baltimore, Maryland." Prepared for the Department of Planning and the Department of Recreation and Parks, June, 1967, 37.

72.   Somerville, "City's Tree Plans Given."

73.   *Baltimore Morning Sun* (15 November 1915).

74. "City Boasts Quarter Million Trees; Fells More In Year Than Is Able To Plant," *Baltimore Evening Sun* (24 May 1955).

75. Furgurson, "City Losing More Trees Than It's Replacing."

76. "Anti-Tree Rebels Prefer Concrete," *Baltimore Evening Sun* (25 April 1967).

77. Ibid.

78. Ibid.

79. "Where the Trees Aren't," *The Evening Sun* (2 May 1967).

80. Evan D. G. Fraser and W. Andrew Kenney, "Cultural Background and Landscape History as factors Affecting Perceptions of the Urban Forest," *Journal of Arboriculture*, Vol. 26, No. 2 (2000): 106–13; John Dwyer, Herbert Schroeder, and Paul Gobster, "The Deep Significance of Urban Trees and Forests," in R. H. Platt et al., eds., *The Ecological City: Preserving and Restoring Urban Biodiversity* (Amherst: University of Massachusetts Press, 1994), 146; and James G. Lewis and Robert Hendricks, "A Brief History of African Americans and Forests," (unpublished collaboration between the Forest History Society and the USDA Forest Service, 2006).

Chapter 6

The chapter epigraph is from the poem, "In a Country Once Forested," in Wendell Berry, *Given: New Poems* (Washington, DC: Shoemaker & Hoard, 2005), 4.

1. Gifford Pinchot, "The Profession of Forestry" (Washington, DC: U.S. Department of Agriculture, 1912), 12.

2. Jack Temple Kirby, *Poquosin: A Study of Rural Landscape and Society* (Chapel Hill: The University of North Carolina Press, 1995), 221.

3. Gary Moll, "Urban Forestry: A National Initiative," in Gordon A. Bradley, ed., *Urban Forest Landscapes: Integrating Multidisciplinary Perspectives* (Seattle: University of Washington Press, 1995), 13.

4. Robert F. Bailey, III, on behalf of the Maryland Department of Natural Resources, *Maryland's Forests and Parks: A Century of Progress* (Charleston, SC: Arcadia Publishing, 2006), 105.

5. Ibid.

6. Author's interview with Offutt Johnson, 6 July 2007.

7.   Author's interview with Francis Zumbrun, 23 August 2007.

8.   Author's interview with Jim Mallow, 21 September, 2007.

9.   Author's interview with Ross Kimmel, 17 July 2007

10. Zumbrun, 23 August 2007.

11. Author's interview with Kirk Rodgers, 30 November 2007.

12. Ibid.

13. Author's interview with Calvin Buikema, 14 August 2007; author's interview with Rebecca Feldberg, 24 November 2007. With regard to the labor force, Guy Hager, Director of Director of Great Parks, Clean Streams & Green Communities at Parks and People, argues that many of the professionals responsible for tree planting are, in fact, really experts at tree removal. To expand the city's tree canopy it will be necessary to rebuild the city's skill base in this vital area. Author's interview with Guy Hager, 16 November 2007. Feldberg concurs: "Bringing on qualified staff is . . . a unique challenge for Baltimore City. We need urban foresters and certified arborists."

14. Cynthia L. Merse, "Historical Geography of Urban Forestry and Roadside Tree Planting in Baltimore" (M.S. thesis, Ohio University, 2005).

15. Author's interview with Mike Galvin, 17 August 2007.

16. Author's interview with William R. Burch, 11 September 2007.

17. Buikema, 14 August 2007.

18. Burch, 11 September 2007.

19. Ibid.

20. "Unsung Heroes we are thankful for in 2001," *Baltimore City Paper* (21 November 2001).

21. Ibid.

22. Mike Galvin, Supervisor of Urban and Community Forestry, Maryland Department of Natural Resources, 17 August 2007.

23. Author's interview with Jackie Carrera, 10 September 2007.

24. Myra Brosius et al., *TreeBaltimore: Doubling Baltimore's Tree Canopy One Tree at a Time*. Draft Urban Forest Management Plan. http://www.ci.baltimore.md.us/government/recnparks/treeBaltimore.html, 49. Accessed 28 October 2007.

25. Galvin, 17 August 2007.

26. Brosius et al., *TreeBaltimore*, 6.

27.   Ibid., 28.

28.   Buikema, 14 August 2007.

29.   Galvin, 17 August 2007.

30.   Author's interview with Guy Hager, 16 November 2007.

31.   Buikema, 14 August 2007.

32.   Author's interview with Gene DeSantis, 21 October 2008.

33.   Brosius et al., *TreeBaltimore*, 42.

34.   Anne Whiston Spirn, *The Granite Garden: Urban Nature and Human Design* (New York: Basic Books, 1984), 179–80.

35.   Burch, 11 September 2007.

36.   Henry W. Lawrence, "Changing Forms and Persistent Values: Historical Perspectives on the Urban Forest," in Gordon A. Bradley, ed., *Urban Forest Landscapes: Integrating Multidisciplinary Perspectives* (Seattle: University of Washington Press, 1995), 35.

37.   See John J. Berger, *Forests Forever: Their Ecology, Restoration, and Protection* (Chicago: Center for American Places at Columbia College Chicago, in association with Forests Forever Foundation, 2008).

Epilogue

1.   *Yale Forest School News*, Vol. 47, No. 4 (October 1959): 75.

2.   Ibid.

3.   "Maxwell to Retire: Parks Head May Get Post In Downtown Tree Plan," *Baltimore Sun* (15 November 1959); and Sasaki, Walker and Associates, Inc., Baltimore Central Business District Master Landscape Plan, February 1961, City of Baltimore Mayor's Street Tree Planting Committee, R. Brooke Maxwell, associated landscape architect.

4.   *Yale Forest School News*, Vol. 49, No. 1 (January 1961): 19.

5.   "R. Brooke Maxwell," *Baltimore Evening Sun* (4 March 1969).

6.   *Yale Forest School News*, Vol. 9, No. 3 (July 1921): 41.

Photography Credits

Page ii. Photographer unknown and photograph undated. Courtesy of the Maryland State Archives.

Fig. 1.1.   Photographer unknown. Courtesy of the Maryland State Archives.

Fig. 1.2.   Photographer unknown and photograph undated. Courtesy of the Maryland State Archives.

Fig. 1.3.   Photographer unknown. Courtesy of the Maryland State Archives.

Fig. 1.4.   Photographer unknown and photograph undated. Courtesy of the Maryland State Archives.

Fig. 1.5.   Photograph by F. W. Besley. Courtesy of the Maryland State Archives.

Fig. 1.6.   Photograph by F. W. Besley. Courtesy of the Maryland State Archives.

Fig. 1.7.   Photograph by F. W. Besley. Courtesy of the Maryland State Archives.

Fig. 1.8.   Photograph by F. W. Besley. Courtesy of the Maryland State Archives.

Fig. 1.9.   Photograph by A. P. Dotsey. Courtesy of the Maryland State Archives.

Fig. 1.10. Undated photograph by W. H. Weaver. Courtesy of the Maryland Historical Society.

Fig. 1.11. Engraving by G. Cooke. Courtesy of the Maryland Historical Society.

Fig. 1.12. Photographer unknown. Courtesy of the Maryland Historical Society.

Fig. 2.1.   Photograph by F. W. Besley. Courtesy of the Maryland State Archives.

Fig. 2.2.   Photographer unknown. Courtesy of the Maryland State Archives.

Fig. 2.3.   Photographer unknown. Courtesy of Kirk P. Rodgers.

Fig. 2.4.   Photograph by F. W. Besley. Courtesy of the Maryland State Archives.

Fig. 2.5.   Photographer unknown. Courtesy of the Maryland State Archives.

Fig. 2.6.   Photographer unknown. Courtesy of the Maryland State Archives.

Fig. 2.7.   Photographer unknown. Courtesy of the Maryland State Archives.

Fig. 2.8.   Photographer unknown and photograph undated. Courtesy of the Maryland State Archives.

Fig. 2.9.   Photographer unknown. Courtesy of the Maryland State Archives.

Fig. 2.10. Photograph by F. W. Besley. Courtesy of the Maryland State Archives.

Fig. 2.11. Photograph by F. W. Besley. Courtesy of the Maryland State Archives.

Fig. 2.12. Photograph by K. E. Pfeiffer. Courtesy of the Maryland State Archives.

Fig. 2.13. Photograph by F. W. Besley. Courtesy of the Maryland State Archives.

Fig. 2.14. Photograph by M. E. Warren. Courtesy of the Maryland State Archives.

Fig. 2.15. Photograph by M. E. Warren. Courtesy of the Maryland State Archives.

Fig. 2.16. Photographer unknown. Courtesy of the Maryland State Archives.

Fig. 2.17. Photographer unknown. Courtesy of the Maryland State Archives.

Fig. 2.18. Photograph by F. W. Besley. Courtesy of the Maryland State Archives.

Fig. 2.19. Photograph by F. W. Besley. Courtesy of the Maryland State Archives.

Fig. 2.20. Photograph by F. W. Besley. Courtesy of the Maryland State Archives.

Fig. 2.21. Photograph by F. W. Besley. Courtesy of the Maryland State Archives.

Fig. 3.1.   Photographer unknown and photograph undated. Courtesy of the Maryland Historical Society.

Fig. 3.2.   Stereoview by W. M. Chase. Courtesy of the Maryland Historical Society.

Fig. 3.3.   Photographer unknown and photograph undated. Courtesy of the Maryland Historical Society.

Fig. 3.4.   Photographer unknown. Courtesy of the Maryland Historical Society.

Fig. 3.5.   Photographer unknown. Courtesy of the Maryland Historical Society.

Fig. 3.6.   Photographer unknown and photograph undated. Courtesy of the Maryland Historical Society.

Fig. 3.7.   Photographer unknown. Courtesy of the Maryland Historical Society.

Fig. 3.8.   Photographer unknown. Courtesy of the Maryland Historical Society.

Fig. 3.9.   Photographer unknown and photograph undated. Courtesy of the Maryland Historical Society.

Fig. 3.10. Photographer unknown. Courtesy of the Maryland Historical Society.

Fig. 3.11. Photographer unknown. Courtesy of the Maryland Historical Society.

Fig. 3.12. Bernet Album. Photograph undated. Courtesy of the Maryland Historical Society.

Fig. 4.1.   Photographer unknown. Courtesy of the Maryland State Archives.

Fig. 4.2. Photographer unknown and photograph undated. Courtesy of the Maryland State Archives.

Fig. 4.3. Photographer unknown. Courtesy of the Maryland State Archives.

Fig. 4.4. Photographer unknown. Courtesy of the Maryland State Archives.

Fig. 4.5. Photographer and cartographer unknown. Courtesy of the Maryland State Archives.

Fig. 4.6. Photographer unknown and photograph undated. Courtesy of the Maryland State Archives.

Fig. 4.7. Photograph by F. W. Besley. Courtesy of the Maryland State Archives.

Fig. 4.8. Photograph by F. W. Besley. Courtesy of the Maryland State Archives.

Fig. 4.9. Photographer unknown and photograph undated. Courtesy of the Maryland State Archives.

Fig. 4.10. Photographer unknown and photograph undated. Courtesy of the Maryland State Archives.

Fig. 4.11. Photograph by F. B. French. Courtesy of the Maryland State Archives.

Fig. 4.12. Photograph by F. W. Besley. Courtesy of the Maryland State Archives.

Fig. 4.13. Photograph by F. W. Besley. Photograph undated. Courtesy of the Maryland State Archives.

Fig. 4.14. Undated photograph by J. J. Chisolm, II. Courtesy of the Maryland State Archives.

Fig. 4.15. Photographer unknown and photograph undated. Courtesy of the Maryland State Archives.

Fig. 4.16. Photographer unknown and photograph undated. Courtesy of the Maryland State Archives.

Fig. 4.17. Photograph by F. W. Besley. Courtesy of the Maryland State Archives.

Fig. 4.18. Photographer unknown and photograph undated. Courtesy of the Maryland Historical Society.

Fig. 4.19. Photographer unknown and photograph undated. Courtesy of the Maryland State Archives.

Fig. 4.20. Photograph by M. E. Warren. Courtesy of the Maryland State Archives.

Fig. 4.21. Photographer unknown. Courtesy of the Maryland State Archives.

Fig. 4.22. Photograph by K. E. Pfeiffer. Courtesy of the Maryland State Archives.

Fig. 4.23. Photograph by F. W. Besley. Courtesy of the Maryland State Archives.

Fig. 4.24. Photograph by M. E. Warren. Courtesy of the Maryland State Archives.

Fig. 4.25. Photograph by M. Hawes. Courtesy of the Maryland State Archives.

Fig. 4.26. Photographer unknown. Courtesy of the Maryland State Archives.

Fig. 4.27. Photograph by F. W. Besley. Courtesy of the Maryland State Archives.

Fig. 4.28. Photograph by A. Aubrey Bodine, *Baltimore Sun*. Courtesy of the Maryland State Archives.

Fig. 4.29. Photographer unknown. Courtesy of the Maryland State Archives.

Fig. 4.30. Photograph by A. Aubrey Bodine, *Baltimore Sun*. Courtesy of the Maryland State Archives.

Fig. 4.31. Photographer unknown. Courtesy of the Maryland State Archives.

Fig. 4.32. Photograph by A. Aubrey Bodine, *Baltimore Sun*. Courtesy of the Maryland State Archives.

Fig. 4.33. Photograph by A. Aubrey Bodine, *Baltimore Sun*. Courtesy of the Maryland State Archives.

Fig. 4.34. Photograph by A. Aubrey Bodine, *Baltimore Sun*. Courtesy of the Maryland State Archives.

Fig. 4.35. Photographer unknown. Courtesy of the Maryland State Archives.

Fig. 4.36. Photograph by A. Aubrey Bodine, *Baltimore Sun*. Courtesy of the Maryland State Archives.

Fig. 4.37. Photograph by A. Aubrey Bodine, *Baltimore Sun*. Courtesy of the Maryland State Archives.

Fig. 4.38. Photograph by S. Proctor Rodgers. Courtesy of Kirk P. Rodgers.

Fig. 5.1.   Photograph by F. W. Besley. Courtesy of the Maryland State Archives.

Fig. 5.2.   Photograph by S. Proctor Rodgers. Courtesy of Kirk P. Rodgers.

Fig. 5.3.   Photograph by S. Proctor Rodgers. Courtesy of Kirk P. Rodgers.

Fig. 5.4.   Photograph by S. Proctor Rodgers. Courtesy of Kirk P. Rodgers.

Fig. 5.5.   Photograph by Lowell Besley. Courtesy of Kirk P. Rodgers.

Fig. 5.6.   Photographer unknown. Courtesy of the Maryland State Archives.

Fig. 6.1.   Photographer unknown and photograph undated. Courtesy of the Maryland State Archives.

Fig. 6.2.   Photograph by the author.

Fig. A.1.   Photograph by K. E. Pfeiffer. Courtesy of the Maryland State Archives.

Fig. A.2. Undated photograph by M. E. Warren. Courtesy of the Maryland State Archives.

Index

About the Author

Geoffrey L. Buckley was born in Washington, DC, in 1965. He received his B.A. in human ecology and American history from Connecticut College, his M.A. in geography from the University of Oregon, and his Ph.D. in geography from the University of Maryland, College Park. His research interests include resource conservation and sustainability; management of public lands, especially state forests and urban green spaces; environmental justice; and the evolution of mining landscapes. Buckley's first book, *Extracting Appalachia: Images of the Consolidation Coal Company*, was published in 2004 (Ohio University Press). He is currently an associate professor in the Department of Geography at Ohio University in Athens, where he lives with his wife, Alexandra, and their three children, Ingrid, Peter, and Owen.

About the Book

America's Conservation Impulse: A Century of Saving Trees in the Old Line State was brought to publication in an edition of 1,000 hardcover copies with the generous financial support of the USDA Forest Service, Ohio University, and the Friends of the Center for American Places, for which the publisher is most grateful. The text was set in Adobe Caslon Pro, and the paper is 70-pound weight. The book was professionally printed and bound in the United States of America.